THE NAKED MAN

Also by Desmond Morris

The Biology of Art
The Mammals: a Guide to the Living Species
Men and Snakes (coauthor)
Men and Apes (coauthor)
Men and Pandas (coauthor)
Zootime
Primate Ethology (editor)
The Naked Ape
The Human Zoo
Patterns of Reproductive Behaviour
Intimate Behaviour
Manwatching: a Field-guide to Human Behaviour
Gestures: Their Origins and Distributions (coauthor)
Animal Days (autobiography)
The Soccer Tribe
The Giant Panda (coauthor)
Inrock (fiction)
The Book of Ages
The Art of Ancient Cyprus
Bodywatching: a Field-guide to the Human Species
Catwatching
Dogwatching
The Secret Surrealist
Catlore
The Human Nestbuilders
Horsewatching
The Animal Contract
Animalwatching: a Field-guide to Animal Behaviour
Babywatching
Christmas Watching
The World of Animals
The Naked Ape Trilogy
The Human Animal: a Personal View of the Human Species
Bodytalk: A World Guide to Gestures
Catworld: a Feline Encyclopedia
The Human Sexes: A Natural History of Man and Woman
Cool Cats: the 100 Cat Breeds of the World
Body Guards: Protective Amulets and Charms
The Naked Ape and Cosmetic Behaviour (coauthor) (in Japanese)
The Naked Eye (autobiography)
Dogs: a Dictionary of Dog Breeds
Peoplewatching
The Silent Language (in Italian)
The Nature of Happiness (in Italian)
The Naked Woman

THE NAKED MAN

A Study of the Male Body

Desmond Morris

Thomas Dunne Books
St. Martin's Press
New York

THOMAS DUNNE BOOKS.
An imprint of St. Martin's Press.

Grillz: Words and Music by Jermaine Dupri, Rich Harrison, James Phillips, Nelly, Paul Michael Slayton, Ali Kenyatta Jones, Cameron F. Gipp, Beyonce Giselle Knowles, Sean Garrett, Kelendria Rowland, Michelle Williams, Clifford Harris and Dwayne Carter © 2005, Shaniah Cymone Music/ EMI April Music Inc/ Dam Rich Music/ EMI Blackwood Music Inc/ Jackie Frost Music/ Basajamba Music/ Paulwall Publishing/ Air Control Music Inc/ Writers Designee/ Kelendria Music/ Christopher Garrett's Publishing/ Hitco South/ Beyonce Publishing/ Point East Music/ Universal Music Inc, USA. Reproduced by permission of EMI Music Publishing Ltd, London W8 5SW © Copyright 2005 Universal Music Publishing Limited (22.5%)/Universal/MCA Music Limited (3.75%)/Sony/ATV Music Publishing (UK) Limited (2.9%). Used by permission of Music Sales Limited. All rights Reserved. International Copyright Secured.

www.thomasdunnebooks.com
www.stmartins.com

Library of Congress Cataloging-in-Publication Data

Morris, Desmond.
    The naked man : a study of the male body / Desmond Morris. — 1st U.S. ed.
        p. cm.
    Includes bibliographical references and index.
    ISBN-13: 978-0-312-38530-9
    ISBN-10: 0-312-38530-7
    1. Men. 2. Men—Physiology. 3. Men—Evolution. 4. Human body—Social aspects. 5. Sex role. I. Title.
HQ1090.M669 2009
306.4—dc22

                                                                2009010687

First published in Great Britain by Jonathan Cape, a division of
The Random House Group Limited

First U.S. Edition: August 2009

10 9 8 7 6 5 4 3 2 1

CONTENTS

INTRODUCTION

The role of women in modern society and their treatment as females in a male-dominated world has been the subject of close examination and debate for many years. The feminist movement of the 1970s gave new focus and direction to this debate with a great deal of research in the next four decades devoted to the human female, including my own book *The Naked Woman* published in 2004. While many wrongs have been righted, at least in the West, across the globe women are still often treated as property and prohibited from sharing in social, economic and political power systems. All major religions have favoured the male over the female and have done so for at least two thousand years. One has to go back a very long way to find the great Mother Goddess and the Earth Mother, and even further, to the tribal societies of primeval times, to find women at the very centre of human society, with the men on the periphery, out hunting for food.

So it is little wonder that authors, examining the male and female gender today, have concentrated mostly on the beleaguered female. Few books have dealt with the male of the species, and investigated his strengths and his weaknesses. He has been viewed as the entrenched enemy, the cause of all social ills, and his special qualities have been largely ignored. As a starting point for reassessing the nature of the human male, this book first looks at his evolutionary success story, and then goes on to study the masculine body from head to toe, looking at each bit of the male anatomy – the eye, the ear, the beard, the chest, and so on – reporting on the biological features in each case and then describing the many ways in which these features have been modified, suppressed or exaggerated by local customs and by changes in social fashions. It is not a medical text – it does not go inside the body. It deals only with the surface features. In a sense, it is a natural history of the Naked Man, a zoological portrait, viewing him as a fascinating specimen of a far from rare, but nevertheless endangered, species.

This book is the natural sequel to *The Naked Woman* and follows the same structure. One crucial difference is that this time I have added a final chapter looking at sexual preferences. This has become a major issue in recent years, one with starkly polarised opinions. In some countries it is now a crime to persecute homosexuals; in others a homosexual act is still punishable by death. It seemed to me that a calmly objective, biological report on this aspect of human behaviour was badly needed and I have done my best to provide one.

In writing this book I have been greatly helped by my wife, Ramona. As with all my books, she has contributed valuable research and editing assistance, but in this case her contribution has been exceptional. I suspect that the subject matter has had something to do with this.

1. The Evolution

No life form has had a greater impact upon this planet than the human male. Explorers, inventors, architects, builders, warriors and foresters have almost always been males and they have changed the surface of the earth to a degree that makes all other species seem insignificant. In the seas they may have had to take second place to the lowly organisms that constructed the vast coral reefs, but on land the human male reigns supreme, both as destroyer of natural features and as constructor of artificial ones. What is it about the human male that has made his legacy so utterly distinct from that of all other life forms, including even the human female? To find the answer we have to return to prehistoric times and take a look at the challenges that faced early men and that helped to mould their unique personality.

When our ancient ancestors descended from the trees, abandoned the vegetable diet of fruits, nuts and roots that was favoured by other monkeys and apes, and turned to hunting and meat-eating as a new way of life, they were at a considerable disadvantage. In taking this dramatic step they placed themselves in direct opposition to mighty predators such as the lions and leopards, hunting dogs and hyenas. Physically, they were no match for these specialised killers. The human body was puny by comparison, lacking sharp claws and fangs. They had to find some other way of competing and it was this pressure that was to fashion the human male in a new mould. They had to use their brains instead of their brawn. The human skull started to swell and intelligence increased.

1

With their bigger brains, prehistoric human hunters could employ cunning beyond anything their rivals could offer. They could not outrun the specialised carnivores, but they could outsmart them. In addition to increased intelligence they needed three other improvements. They had to modify their competitiveness, tempering it with increasing cooperation, so that they could work as an active team. They had to become more inventive, so that they could develop novel techniques. And they had to stand up on their hind legs, so that they could free their front feet from the toil of walking and running, and develop them into grasping hands that could fashion and refine implements and weapons.

With these improvements, the early bands of tribal hunters became a formidable force on the land. They could drive the great carnivores off their kills and scavenge the meat. Or they could make the kills themselves, devising manoeuvres, ambushes and traps that defeated even the most powerful of prey species.

With these improved skills came a new challenge. There was suddenly so much to eat that the leading hunters could not consume it all themselves. There was enough for the whole tribe and food-sharing became a basic facet of human society. Today we take this very much for granted, but to a monkey in a tree it is an alien concept. Animals that primarily eat vegetables never share them with their companions. Each vegetable-eater gobbles up what he or she finds. Vegetable-eating is always a selfish act. But if there is a food surplus, as there is after a big kill, the whole tribe can eat. The human feast was born.

As the efficiency of the human hunters gradually improved, their personality started to change. They became more and more distinct from the human female in mental outlook and physical build. Hunting was a dangerous pursuit and the females of the early tribes were too valuable reproductively to be risked in the chase. They became specialised as cautious, caring, maternally efficient individuals, operating at the centre of society and dealing with many different tasks in primitive settlements, while the more expendable males became more and more adventurous and increasingly roamed abroad in search of prey, taking risks of a kind that no tree-dwelling, fruit-eating monkey would ever contemplate.

2

The male brain and the male body both underwent special modifications during this crucial phase of human evolution. Mentally, the hunter not only became more venturesome, more cunning and more cooperative, but also more single-minded and more persistent, capable of planning long-term strategies as well as short-term tactics. Physically, the male body had to become increasingly muscular and athletic, and had to sacrifice the precious fat deposits that gave the tribal females their more curvaceous forms and also provided them with vital nutritional reserves during the inevitable, occasional periods of food shortage.

In this way the primeval human male evolved into a startlingly efficient prey-killer and early human tribes began to multiply and spread rapidly all over the globe. This condition, the classic hunter/gatherer society, lasted for hundreds of thousands of years, until a new phase appeared about 10,000 years ago with the arrival of farming. It began when our ancestors started to improve their gathering techniques. Instead of searching for their vegetable foods, early agriculturalists began to plant and grow certain crops near their settlements, attracting herbivorous prey animals. Instead of chasing after their prey, the hunters could get the prey to come to them. They began penning the animals in and keeping them as captive meat. Now they could enjoy a feast whenever they felt like it.

As the captive prey animals began to breed, these first farmers realised that they could now control the reproduction of their prey, and possess their own livestock. The agricultural revolution was upon us. In relative terms, it all happened so quickly that the personality of the male hunters did not have time to evolve to match the changing circumstances. In their genes they remained mighty hunters, while in their daily lives they became farmhands and herders, crop-sowers and harvesters. The drama of the hunt and the thrill of the chase became the drudgery of the farmyard. The huge advantage of having a food surplus was tempered by the loss of adventure and the bold spirit of the primeval hunting pack.

How did the new farming male cope with this loss? Sport-hunting was developed as a way of reliving the excitement of the kill, but

3

this was not enough. The human male had been forged by evolution as a tough, inventive, cooperative, risk-taking being and needed to find some way of expressing himself that did not deny his biological heritage. In fact, he found two – one destructive and one creative.

The destructive one was warfare, in which rival males were treated as prey to be hunted down and killed. This gave the risk-hungry male all the dangers he could dream of and, as weapons became more and more sophisticated, far more dangerous than he had ever envisaged. If these two corrupted forms of hunting, sport-hunting and war-hunting, had been the only two ways in which the human male had faced the mental challenge of the farming era, our species would have remained in a sorry state. But there was also a constructive response to the loss of food-hunting. The ability to concentrate on long-term goals that was an integral part of the primeval chase was co-opted by Neolithic males and put to work in the service of major new endeavours.

At first, progress was painfully slow, but, as the centuries passed, flimsy huts became great buildings; crude body-painting became great art; simple tool-use became refined craftsmanship. Villages grew into towns and cities, technical specialisations blossomed and all the complexities of modern civilisation began to beckon.

The male inventor was the new, improved risk-taker, forever seeking novelty. In this role he was set firmly against the destructive male and, although the two still continue to reflect twin faces of the human male – the breaker and the maker – our modern world is a living testimony to the fact that creative male energy has managed largely to overshadow the negative one.

In recent times there has been much talk about the 'Redundant Male', the suggestion being that with new, artificial fertilisation techniques men will soon become obsolete. This theory first became popular in the 1970s when leaders of the feminist movement announced that clitoral orgasms were superior to vaginal ones and that males were not worth the trouble in the bedroom. However, even if men were not necessary for sexual pleasure, there was still the tricky problem of how to procreate the next generation of feminists. A few champion ejaculators would have to be kept for this purpose, with sperm samples being ordered whenever they were needed.

Since then, advances in reproductive technology have been made that suggest that one day, in the not too distant future, even sperm will not be necessary. Women will be able to have their eggs fertilised in the laboratory without any male element being involved, and then have them reinserted in the uterus to grow into a new generation of females. Lesbian pairs will be formed to create a new type of family unit with baby girls being reared in a male-free world.

According to this ideal, the absence of males will mean an end to warfare, testosterone-fuelled violence, aggressive sports, football hooligans, political extremists, rapists, religious terrorists and all the other destructive aspects of the masculine world. In its place will be the caring, sharing, gentler, more intelligent world of the human female. Quiet common sense will replace savage conflicts of honour, and life will become a warm, safe, friendly experience rather than a cruel, anxiety-ridden ordeal.

How all the existing men would be disposed of is not clear. Perhaps they are just ignored and allowed to grow old until the male gender slowly fades away. Or perhaps they would be massacred, as proposed in the manifesto of the radical feminist movement called SCUM (The Society For Cutting Up Men). Eventually they will be no more than a distant memory, and a testosterone-free planet will rotate to the sound of female laughter.

It is worth pointing out, on a serious note, that in addition to ridding the world of the destructive elements in the masculine psyche, this extreme scenario would also remove all the constructive elements. There would be far fewer major inventions – they would be considered too risky. There would be far fewer single-minded, long-term projects – too time-consuming when set against the demands of family and daily social life.

If women have always been more sensible than men, men have always been more playful than women. And it is this adult play-fulness that has given the human species many of its greatest achievements.

5

If we allowed the champion of all things male to offer his riposte to the feminist position, he would probably say:

Yes, there may have been great female artists, scientists, politicians, religious leaders, philosophers, inventors, engineers and architects. But for every one of them there have been a hundred men, possibly a thousand. Greatness seems to demand the sort of stubborn perversity that is a predominantly male quality.

It has often been argued that this has been a matter of opportunity – that women were not allowed to develop their true potential. But in practical terms this simply means that women were not great enough to demand that their greatness be recognised. Greatness has to be achieved, not merely postulated, and it is the men who have been driven on by their genetically installed ambitions actually to take the great steps necessary to build our towering civilisations.

Both these extreme views are exaggerations and represent the enormous waste of energy that has gone into the so-called battle of the sexes. The truth is that the human male and the human female make a perfect evolutionary team. They are different in several important ways that have evolved over hundreds of thousands of years to refine the division of labour that exists in the human tribe, but at the same time they are equal in importance. Different but equal, that is the key. The male brain has become specialised in single-minded determination. The female brain has become adept at multi-tasking. The male has become specialised in planning, innovating, risk-taking, spatial problem-solving and muscular expression. The female has become specialised in verbal fluency, and with better developed senses of hearing, smell and touch, and a greater resistance to disease.

Sexually, the human male has changed dramatically from his monkey and ape relatives. While they usually only have one sexual strategy, he has two. The first is to fall in love and form a pair-bond with a particular female. This is not, as is sometimes claimed, a cultural sophistication, but a deep-seated biological quality. The emotional upheaval that accompanies the process of pair-formation is anything but sophisticated. It is deeply physiological with dramatic chemical changes taking place inside the male (and female) body. And it occurs globally, even in those societies that have tried to impose other, inappropriate mating systems on human adults. There are many cases where, once the bonding process has taken place, the couple in question will face imprisonment, torture and even death, rather than abandon their chosen partner.

In evolutionary terms the advantage of human pair-formation is that, in a small tribe, it shares out the females between the male hunters. Unlike other male primates, human males had to cooperate on the hunt. Alone, the individual hunter was not fast enough or strong enough to defeat his prey. The leading hunter needed the active cooperation of his male companions. If the leading male kept all the females in the tribe for himself, he could hardly expect to be given active assistance by the other males when they were out together on the hunt. The pairing system, back at the tribal settlement, created a greater equality between the males. There would still be a social hierarchy, with more dominant and less dominant individuals, but the gradient between the top and the bottom of the hunting pack would be far less steep.

It would be wrong to think of this newly developed cooperativeness as one of our 'refined' qualities, inspired by a new, spiritual sense of acquired restraint and saintly unselfishness. In reality it is one of our most basic animal properties. Moralists often seem to think that, biologically, the human species is competitive and selfish, and that only moral teaching can induce us to behave in a helpful, unselfish, altruistic way. But the truth is that such behaviour is in our genes. If we had not changed genetically to become more helpful towards one another, early human tribes simply would not have survived. Paradoxically, behaving unselfishly was a selfish act.

There was a second, important advantage in the pair-formation system of tribal mating. It created a family unit in which the children knew their father as well as their mother. And it added paternal care to the maternal care found in other species. At a stroke, it doubled the parental protection that growing children received. Powerful paternal feelings are unleashed the moment a human father holds his new baby in his arms and in the years ahead he will devote a great deal of time and attention to the rearing of his offspring.

The reason evolution has changed the human male in this way and converted him into a good father is that our species has such a heavy parental burden that one adult, the mother, cannot cope alone. A female monkey has a much easier task. Her baby is able, from birth, to cling actively on to her fur and ride on her body as she moves about. The newborn monkey is so much more advanced that it does not have to be carried or placed in a nest. It grows rapidly and is soon scampering about near its mother's body, returning quickly to her if danger threatens. Before the mother has another baby to care for, her first is already more or less independent. So she never has a big family to worry about.

In stark contrast, the human mother has a serial litter to care for. Her babies are helpless at birth, demanding endless attention during the early months, and they are still completely dependent on her a year or two later, when the next baby arrives. And so on, until the mother is caring for a whole brood. Having the support of a protective, loving father during this process provides a major boost to the survival chances of the young ones. From the male's point of view, the stronger his paternal feelings the more chance he has of seeing his genetic progeny thrive.

In this respect the human male is more like a bird than a monkey. Birds have helpless offspring too, in the shape of eggs, and the parental burden of incubating them requires the efforts of both the male and the female parent. If the male bird did not share the duties of sitting on the eggs, the female bird would starve to death before she could hatch them. If, to prevent this, she took off to feed when there was no male to take over from her, the uncovered eggs would soon chill and the chicks inside them would die. So pair-bonding in birds is the typical mating system, found in almost every species, and for the same reason it is found in humans, namely the need for intensive parental care.

I said earlier that the human male has not one, but two sexual strategies. The first, as we have seen, is to devote a huge amount of time and effort to his family unit, ensuring that his children have the very best chance of survival. The second is the more primitive one of scattering his seed wherever and whenever he gets the chance. If he finds himself in the company of an adult female who is not his family partner, he may feel the urge to engage in a brief bout of sexual activity with her, even if he is never going to encounter her again. If she produces an offspring as a result of this brief encounter, he will not take any part in its rearing and may not even know of its existence. Lacking his paternal care, the baby will have less chance of survival than one inside a carefully protected family unit, but it will not be without any chance. Furthermore, in instances where the female concerned is already paired with another male, her permanent male partner may believe that the offspring is his and give it his full protection. In this way it will have an excellent chance of survival.

The inevitable question arises as to why a paired human female should take the risk of mating with a strange male when she has a permanent partner available to her to make her pregnant. If she is discovered it could clearly cause serious damage to the stability of her family unit, and yet it does still happen. The reason appears to be that the human female is programmed to assess human males in two different ways. In one assessment, she rates them according to their supportive qualities. She senses how well they will look after her and her offspring, and how socially successful and reliable they are. In the other assessment, she rates them according to their physical fitness. Do their bodies look as though they will pass on good genes to her offspring? In an ideal relationship, the female's permanent partner will be both reliably supportive and also physically impressive, and she will have no genetic reason to stray. But if she has chosen a partner primarily for qualities of protection and caring, then she may be tempted, from time to time, to engage in a little risky sexual activity outside the family unit.

In earlier epochs it was never possible to be certain about how much of this extra-marital activity was going on. In some cultures males went to great lengths to ensure that their female partners had no chance to encounter strange males by keeping them inside the family home for most of the time and by always accompanying them or having them chaperoned when they did allow them out. In some instances custom went even further, with women being forced to cover their bodies completely when they went outside the home. Female circumcision, the surgical removal of the external genitals of young girls, was also practised. This reduced their chances of gaining sexual pleasure, and in that way further reduced their interest in other men. This practice still continues in many parts of the world with at least ten million women alive today who have been sexually mutilated in this way.

Today, in the West, efficient DNA testing has at last made it possible to ascertain, with some accuracy, how many children are the result of the human male's pair-bonding strategy and how many are the result of his more ancient seed-scattering strategy. The results have been surprising. Most married men, in modern times, would have imagined that a family-unit child that had a stranger as a father would be a great rarity. But this has not been proved to be the case.

DNA paternity testing was first introduced during 1995/6, as a means of resolving paternity disputes without the need to go to court. Figures obtained over a seven-year period from 1998 to 2004 revealed that, in Great Britain, 16 out of every 100 children tested for paternity proved to have biological fathers who were not the 'husbands' who were rearing them as their own. Similar tests in Northern Ireland gave almost identical results, differing by only 0.2 per cent. These were much higher figures than anyone had expected.

In the United States, a paternity authority has been quoted as saying: 'The generic number used by us is ten per cent.' In Germany, the Max Planck Institute stated: 'The rate of wrongful paternity in "stable monogamous marriages," ranges from one in ten with the first child to one in four with the fourth.'

A broader study of nine major regions (the UK, US, Europe, Russia, Canada, South Africa, South America, New Zealand and Mexico) revealed that estimates varied from 1 to 30 per cent. This huge variation suggested that there might be something wrong with the samples being used. The problem seemed to arise from the fact that, in many of the reports, the figures were based on cases where true paternity was already being questioned, and where there was some doubt about who was the father. If all those studies were ignored, then the figure was much lower. The investigation concluded,

'The remaining research showed an average paternal discrepancy of 3.7 per cent, or a little less than one in 25.'

To return to the human male's two mating strategies, this means that, even within a society that is, in comparison to earlier times, a sexually liberated society, 24 out of every 25 children are the result of the pair-bonding strategy and only one is the outcome of the seed-scattering strategy. It is clear from this that, despite his well-documented urge for philandering, the human male is essentially a pair-forming being.

How, then, can one explain the male's endlessly roving eye? He may not father too many extra-mural children, but this does not mean that he is as faithful to his mate as a strict pair-bond should imply. The answer comes from an evolutionary trend that has seen the human species become increasingly childlike during the past million years or so. The value of this trend, called neoteny, has been that it has resulted in human beings retaining their childlike sense of playfulness and curiosity well into adulthood. This has enabled them to become increasingly innovative, leading to all the ingenious inventions that have given us our complex modern technologies. But at the same time it has led to this heightened level of curiosity spilling over into other aspects of life, including our most basic animal activities.

With food and drink this trait is not a problem; the result is simply gourmet eating and fine wines. But where sex is concerned it has often played havoc with our primary reproductive strategy. When an already paired male sees an attractive stranger of the opposite sex, his curiosity makes him wonder what it would be like to enjoy her sexually. In the vast majority of cases he is able to keep his curiosity operating at the level of sexual fantasy, but he also occasionally goes further. Usually, once he has satisfied his curiosity, that is the end of the matter, but in many instances there is a disruption of the original pair-bond and often the formation of a new one. This inevitably reduces the quality of his paternal caring for the children of his original relationship, no matter how he tries to repair the damage.

Major family disruptions of this kind were more difficult in the small tribal communities where male reproductive patterns evolved.

But modern society is more complex and the opportunities for pair-bond disruption so much greater that the ancient system is coming under increasing strain in modern times. The divorce rate has increased dramatically and, although some figures quoted are wild exaggerations, it seems that, in twenty-first century America, for example, of those who marry, 34 per cent will experience divorce. A comparable figure for the UK is 36 per cent. So it seems that about one-third of modern pair-bonds collapse, a fact used by professional anxiety-makers as a sign that society is in decay. Another way of looking at the situation, however, is that, even in the midst of modern decadence and liberal sexual attitudes, two-thirds of modern couples do manage to make their pair-bonds work. Given the highly unnatural structure of urban society, to which the human tribal animal has had to adapt, this is a remarkable testimony to the tenacity of the pair-bonding mating strategy.

One argument that is sometimes heard is that, if pair-bonding is such a basic feature of the human species, why is it not total? If it was so valuable to the early tribal groups, why did evolution not make it permanent? There are stories about birds that pair for life with such an intensity of attachment that when one of them dies, the surviving partner never takes a new mate. Why did evolution not develop this extreme mechanism for humans and avoid all the misery and inefficiency of marriage collapse?

The answer, it would seem, is that in primeval times when the new mating strategy was developing in the hunter/gatherer tribes, the males were facing serious dangers in their search for prey and the females were facing difficult births due to the new vertical posture of our species. Either member of a mated pair could easily die young and this would leave the surviving partner reproductively stranded if the pairing mechanism was too rigid. If, on the other hand, after a period of distress and mourning, the surviving young adults could find it possible to form new pair-bonds, then the reproductive rate of the very small tribes could benefit. So, an almost perfect pair-bond would be better, from a survival point of view, than a perfect one.

In evolutionary terms, then, the human male is programmed to form a long-term relationship with a female partner, but with the reproductively valuable possibility that, should she die, he can, after a while, form a new pair. It is this slight weakening of the bond of attachment, so useful in early days, that has been magnified into a serious problem in modern times. The main reason for this has been the way in which the hunting pattern of the male has changed. Instead of setting out on a gruelling, dangerous pursuit of animal prey, he now sets off for the city to engage in a different kind of hunt. Once there, he will find himself in the company of many attractive young women, women who were totally absent from the primeval hunting fields. There were few temptations out on the savannah, but in the big city they are all around him. And this is where his less than perfect tendency to remain exclusively bonded to his mate lets him down.

But whatever its faults and its weaknesses, the fact remains that the majority of human males, in the majority of human cultures, settle in to long-term family units. And despite all the conflicts and the disruptions, these units have proved themselves to be remarkably successful at child-rearing. The proof of this is that the global human population has more than doubled in the past forty years, from 3,000 million to more than 6,000 million today.

If, on a global scale, the human male succeeds as a father, what of his role as a tribal hunter? As already mentioned, his ancient urge to hunt down prey animals has not been lost – evolution works too slowly for that to have happened – but it has been transformed in several ways, so that the hunt has become a symbolic one. All modern competitive sports, for example, are symbolic forms of hunting. They are all concerned with being skilled at either chasing or aiming, or both, and chasing and aiming are, of course, the basic ingredients of the primeval hunt. A huge industry has grown up around this symbolic activity, with enormous prizes for the champion chasers (like Michael Schumacher) or aimers (like Tiger Woods), and with vast throngs of would-be champions cheering them on from the sidelines. Team sports like football and basketball, to name but two, involve both chasing and aiming and, in addition, provide the modern sportsmen with the same sort of involvement in planning, cooperation, tactics and strategies that so absorbed their ancient counterparts.

13

It is noticeable that sport has no end product. It manufactures nothing. Its end point consists of a victorious champion holding aloft the precious prize, the symbolic prey. This is usually a cup, a statue or a plaque of some kind, an inedible, useless piece of metal. There may be a celebratory feast at the end of a major contest, just as there was at the end of a successful hunt, but the sporting activity itself is not productive in any way. It simply satisfies a deep-seated urge in the human male to re-enact, either as performer or onlooker, the skilful, physical challenge of the hunt. Every time the dart hits the bull's eye, the billiard ball sinks into a pocket, the puck crosses the goal-line, the baseball goes for a home run, the cricket ball is hit for six, the football strikes the back of the net, and so on, the modern hunter and his followers let out a roar of triumph that, just for a moment, is as primeval as Tarzan's great cry in the jungle. The moment of aiming has succeeded. The tribe will flourish. It is little wonder that, despite a great deal of female interest, the world of sport remains a predominantly male domain.

One of the special attributes of the human male is his physical strength. His athletic body, required for the hunt, developed advanced musculature that set him apart from the human female. On average, a man is 30 per cent stronger than a woman. His body contains 26 kg (57 lb) of muscle; hers contains only 15 kg (33 lb). In complete contrast, only 12.5 per cent of the male's body weight is fat, whereas the figure for the female is 25 per cent. Nowhere is the division of labour in human evolution more clear than this. This is underlined by the fact that, when women indulge in muscle-building, they start to look more and more like men. It is impossible for them to develop a muscle-bound body that is feminine in appearance. If they go to the extreme of high-level competition in body-building, they even cease to ovulate.

The design of the male body means that he can lift much greater weights than the female – again an important quality when the hunters were carrying home the kill. The average male can lift twice his body weight; the average female can lift only half her weight. Internally, the male organs, the heart, lungs and bones, are bigger than those of the female, providing a valuable support system for the stronger male's muscles. And the male's blood contains more haemoglobin than the female's.

The whole skeleton of the male body is bigger, giving the muscles a more powerful base from which to operate. And the average male is about 10 per cent heavier and 7 per cent taller than his female counterpart. The development of this more powerful male body can be detected as early as day one. The newborn male baby is, on average, both longer and heavier than the average female baby, and it displays more vigorous limb movements. It also has a higher basal metabolism and retains this all through life. Even at birth the male is showing signs of a greater athleticism than the female.

Among children, males display a greater visual acuity, something that will be useful on the hunt when they become adult. When at play, boys typically indulge in power play, with far more pushing, shoving, running, jumping, hammering and banging than is seen with girls. And their level of curiosity is already ahead of that of girls, revealing the birth of what will later become adult male risk-taking.

It has become fashionable in recent years, as part of the creed of sexual equality, to suggest that these differences between small boys and small girls are not inborn, and that any divergence we see is entirely the result of adults imposing artificial male and female roles upon them. Any careful study of toddlers soon dispels this idea. The signs are present very early on, in any playgroup, where no parental bias is operating, and even before the parents realise that the differences exist.

In fact, the idea of childhood sameness with boys and girls is not only wishful thinking, carefully tailored to fit a theory, but it also happens to be quite unnecessary for that particular theory. The idea of sexual equality is, in itself, perfectly suited to our species, but it does not have to depend on males and females being the same (except for a few minor anatomical differences, of course). The important division of labour that evolved in early human tribes did not create a dominant gender and a subordinate gender, but two genders that were urgently dependent on one another and that had equal importance. The fact that boys have special inborn differences from girls does nothing to damage the concept of equality of importance.

Unfortunately for today's male, the conditions of modern civilisation have rendered his muscular superiority obsolete in the majority of tasks. Sitting behind an office desk or a factory bench, standing behind a shop counter, or slumped in front of a computer screen, requires little athleticism. In physical terms these are not fitting occupations for the human male body. It needs more active expression than this, if it is to be faithful to its tribal roots.

Some men respond to this problem by setting aside time to engage in 'keep fit' activities of one kind or another, but the majority can't be bothered. For a few men, however, the urge to demonstrate their physical maleness becomes overpowering and they undertake exceptional feats of endurance and compete in dramatic displays of body strength. Everything from weight-lifting to mountain-climbing, and from arm-wrestling to polar trekking is undertaken, not because it will serve any practical purpose but because it will enable certain males to show their disgust at the increasing softness of twenty-first-century men. To the vast majority, these displays seem pointless and a waste of energy, but they are nevertheless a vivid reminder of what has been lost by modern men.

As television and computers have made life increasingly sedentary for the human male, we have also seen the arrival of dangerous sports clubs and other organisations that cater to the young male who is determined to engage in high-risk, extreme forms of physical activity. Some of the most popular of these new sports are:

Base-jumping, which involves leaping off the top of a tall structure such as a skyscraper, a bridge or an electrical tower, wearing a parachute and hoping it will open in time.

Cave-diving, the exploration of labyrinthine underwater systems, wearing diving equipment, but with the risk of getting completely lost hundreds of feet below the surface and then running out of air.

Speed-skiing, in which participants travel downhill at up to 160 miles an hour, wearing aerodynamic suits and specialised skis. One crash in these events usually results in death.

Supercross, in which riders fly through the air on a motorcycle while doing back-flips and other aerobatic manoeuvres.

If the recent spreading popularity of these and many other extreme sports does not demonstrate the deep-seated male urge to take serious risks, nothing will.

For some males, physical challenges have little appeal, but they still want to enjoy the thrill of risk-taking. For them, there is a whole range of possibilities, from playing the stock market to gambling at Las Vegas. Life in the financial centres of the great cities of the world involves taking risks almost every minute of the day and it is no accident that this is a largely masculine pursuit. Serious gambling, too, is a male-dominated pursuit. It is true that, if you visit the casinos of Las Vegas, you will also see many women, but the majority of them are playing the slot machines. The high-rollers around the poker tables are mostly men. It is the same at local bingo games and race tracks, with low stakes for women and high stakes for men. Women are, by nature, sensible and cautious; men are, according to your point of view, either brave or stupid.

Physical exertion and risk-taking are only two aspects of the primeval hunt that men today still feel impelled to re-create in some symbolic form. There is also the urge to bring home the kill. Shopping for meat in the local supermarket does not quite do the trick. Something more is needed. One answer is to become obsessed with collecting things. The collector, again usually male, develops a passion for some particular category of object and then sets about amassing as many good examples as he can. The category he selects can be almost anything, from Old Master paintings to match-box labels. It really doesn't matter, just as long as there are many to choose from, with collections usually starting out with the common examples and then progressing to the increasingly rare ones. Tracking these down and bringing them back home to add to the growing collection becomes the special joy of the object-hunter.

The intensity of object-hunting can become so great that it virtually takes over a man's life. Some houses are so stuffed full of collected items that it is almost impossible to find an inch of space to spare. And some of the objects collected are truly bizarre. They include such unlikely things as lawnmowers, dog collars, air-sickness bags, vintage vacuum cleaners, electric toasters, and quack medical devices. Almost anything is fair game, once the hunter's prey is no longer fresh meat but has become a symbolic substitute.

At the top end of the scale, the atmosphere in those very special hunting grounds, the world's great auction houses, has to be felt to be believed. When rival hunters are in full cry, admiring gasps can be heard in the sale rooms as prices go through the roof. Picasso's *Boy with a Pipe*, which was sold for $104 million in May 2004, holds the world record for a painting sold at auction. Even this was outdone recently by a private purchase arranged through Sotheby's of a painting by Jackson Pollack that fetched a world record price of $140 million.

Another characteristic of the early hunting males was that, when they were resting from the chase, they would spend hours cleaning, repairing, caring for and generally improving their weapons, the vital implements that stood between them and starvation. This fascination for primitive technology is today reflected in a masculine love of gadgets and machines, instruments and equipment, a pursuit that appeals to very few women.

After a successful hunt, there was the inevitable celebration and the time to tell tall tales about the dangers experienced on the chase. In modern times this has been transformed into hard-drinking sessions for groups of males after work. Anthropologists have described this as the 'separation of drinking from the female-dominated domestic arena as a means of constructing masculinity'. In other words, by getting together and downing generous amounts of alcohol, modern males are able to re-create for themselves, momentarily, the sensation of being part of a loyal hunting pack.

It is important, on these occasions, to buy one's round and to hold one's liquor. Anyone failing on either of these counts loses status in the group. In other words they must demonstrate both that they are sharing and that they are tough. In various countries at different times, these drinking sessions have become formalised. Mens' clubs with a strict membership have been formed and drinking games and other rituals have been invented to give the drinking sessions a heavier significance. In some countries, other narcotics are employed in place of the usual alcohol. In Yemen, for instance, male groups gather every day to chew *qat* (pronounced *gat*), the leaves of a narcotic plant. To have any social significance at all, a Yemeni male must belong to one of these *qat* gatherings, from which all women are totally excluded.

Some kind of game is often employed as a focus to bring men together in a mildly competitive way. The oldest one known to mankind, and one that was almost certainly used by the primitive hunters themselves, is the African board game called *mancala*. It has been played for at least 3,400 years, and probably much longer. Its advantage is that it can be played even without special tokens or wooden boards. Small pebbles and holes scooped in the dry earth are sufficient to get a game going, as the hunters squat together, relaxing after a chase. There are many other masculine games, such as *boules* in France, chess in Russia, darts in Britain, poker in America, and so on. Each of these beckons the wedded man away from the bosom of his family and briefly back into the company of the all-male gang.

In some countries all-male activities are much closer to the original hunting pattern. A group of men will take off for the woods to camp, fish, trek, or carry out some other such pseudo-primitive pursuit. Others will go on safari to Africa, or explore ancient ruins in Central America.

Time and again, as one moves around the world, it is possible to discover all-male pastimes of this sort, each with its local, official reasons for existing. But underlying all of them is the re-enactment of the social bonding of the primeval hunting party, the bonding that is necessary to ensure unquestioning support at sudden moments of crisis.

The twenty-first-century hunter can fulfil his urge to chase and aim by engaging in sport. He can express his physical strength by engaging in fitness routines. He can demonstrate his bravery and his risk-taking skills, from the physical to the mental, from dangerous pursuits to gambling. He can satisfy his desire to bring home the kill by searching for rare objects and assembling a collection of them. He can display his urge to care for and improve his weapons, by developing a technical skill of some kind. And finally, he can relive the successful hunter's feast by engaging in bouts of social drinking. The ancient, primeval hunter may be dead, but the modern, symbolic hunter lives on.

2. THE HAIR

The hair that grows on top of a man's head is one of the strangest features of his body. Imagine what it must have been like for our ancient ancestors before there were brushes and combs, scissors and knives, hats and clothes. For more than a million years they were running around with almost naked bodies topped by a great mass of overgrown fur. While the hair on their trunks and limbs shrivelled to insignificance and exposed the whole of that skin surface to the open air, the hair on their scalps sprouted into a huge woolly bush or a long swishing cape. Unadorned and unstyled they must have looked amazing to other primates. What manner of ape was this?

Inherent in this question is the solution to the puzzle as to why the human species has such unusual scalp hair. The answer is that it made us look very different from other primate species. It was our species signal, visible from afar. Great bushy heads on top of smooth naked bodies identified us immediately as human. Our extravagant tresses were carried like a flag.

It is easy to forget this today because scalp hair has been co-opted by fashion as an extra gender signal. In almost every culture, males and females dress their hair in masculine and feminine styles. So all-pervasive is this trend that we can be forgiven for overlooking the fact that, before the onset of balding in males, the structure of male and female scalp hair is identical. We do, of course, have hair gender signals – moustaches, beards, hairy chests and the rest; but on top of the head, where our hair growth is at its most luxuriant, we enjoy complete sexual equality, both as children and as young adults. A flamboyant head of hair is neither feminine nor masculine, it is *human* and it set us apart from our primate relatives as we evolved into a distinct species.

If this seems improbable, it is only necessary to look at the hair patterns of various monkeys and apes. Closely related species often show dramatic differences in the colour, shape and length of their head hairs. Some have coloured caps of hair, strikingly differentiated from the rest of the head hair; others display long moustaches. Still others sport impressive beards; some are even bald. Clearly, the tendency to employ head hair differences as specific labels is commonplace among primates, so it should not be too surprising that our own species employed a similar identification mechanism. What *is* surprising is that we carried it so far. Our scalp hair display became so exaggerated that as soon as we were sufficiently advanced, technologically, to attack it with knives and scissors we set about curtailing it in a thousand ways. We snipped it and cut it, shaved it and tied it, braided it and twisted it, and stuffed it away under hats and caps. It was as though we could not stand the encumbrance of our primeval hair pattern any longer. The great weight of our sprouting locks had to be lightened in some way.

It would be wrong to conclude from this that we became anti-hair. What happened was that we made a cunning exchange of old hair for new. In our primeval condition, the human scalp pattern relied on sheer bulk to label our species. This made us highly conspicuously different, so that it was an efficient label, but it also happened to be rather cumbersome. When our species advanced to the stage where we were able to style our hair, we found many new ways of making it conspicuous, with strange shapes, colours and adornments, while at the same time reducing its bulk. In modern times, with the whole paraphernalia of hats and wigs and hair dressing, we could have the best of all worlds, having functional, working heads one moment and then fanciful display heads the next.

Before examining these new trends, it is important to take a closer look at the raw material on which they are based, at the natural hair itself. On average, each human scalp boasts about 100,000 hairs. Fair-haired people have more individual hairs than darker people. A blond-haired person has about 140,000 hairs; one with brown hair, 108,000; but redheads are unlikely to have more than a mere 90,000.

Each hair grows from a small skin pocket called a follicle that has at its base a papilla. This minute lump of tissue is the hair-maker. It is rich in blood vessels and it supplies the raw materials that are converted into the hair cells. These cells keep forming on the surface of the papilla, the new ones pushing up the old ones, making the hair grow in length. Eventually the hair root, the part beneath the skin surface, becomes so long that the tip of the hair emerges from its little pocket. As it does so it becomes hardened. The visible section of the hair, that gets longer and longer, is called the hair shaft. The shaft increases in length by about a third of a millimetre a day.

The speed at which hair grows varies with age and health. It grows most slowly in old age, illness, pregnancy and cold weather. It grows quickest during convalescence after a serious illness, apparently as a compensation mechanism after a period of impeded growth. In healthy people it grows fastest when we are between the ages of sixteen and twenty-four. During this period the annual increase is up to 7 inches (17.7 cm) a year, compared with an overall average of about 5 inches (12. 7 cm).

The life span of an individual hair is about six years, which means that, for a healthy young adult, an untrimmed hair would grow to roughly 42 inches (106.6 cm) in length, before falling out and being replaced. This means that straight-haired young men and women who never cut their hair would have flowing locks reaching to their knees.

Apart from its inordinate length, another oddity of human hair is that it does not show a seasonal moult, as with many other animals. We could easily step up our hair loss at the start of the summer and slow it down in winter, to create a more finely tuned insulation layer, but we show no signs of this. Each individual hair has an independent life cycle. At any one moment, 90 per cent of scalp hairs are actively growing and 10 per cent are resting. The resting ones are scattered throughout the others and they remain in this inactive state for about three months before dropping out. This means that we each lose between 50 and 100 head hairs every day.

When a hair drops out, both the long shaft and the short root become detached, but the tiny papilla at the base of the follicle stays put. This little bud then starts to sprout a new hair to replace the old one. It remains active for another six years, when it once again stops growing new cells and becomes dormant. After another three-month resting period it sheds its old hair and repeats the whole process. With our extended lifespan today, each papilla can be expected to repeat its life cycle about twelve times, creating twelve complete hairs, one after the other, each hair being several feet in length. It follows from this that if an individual human being lacked the dormant phase in the hair cycle, he or she would be able to grow locks of hair up to 30 feet in length. This freak condition does seem to have occurred at least once. Swami Pandarasannadhi, an Indian monk from a monastery near Madras, was reported to have unkempt hair that stretched for 26 feet (7.49 m).

The opposite effect, a permanent switching-off of hair growth, is much more common. It never occurs during childhood, but when sexual maturity arrives strange things start to happen on top of the male head. The male hormones flooding through the system de-activate certain selected hair papillae. The ones around the sides of the head are spared, but those on the crown of the head are knocked out of action. As individual hairs fall out they are not replaced. The papillae, instead of becoming dormant for three months and then starting up again, become dormant forever. As a result their owner becomes bald.

Balding is usually a gradual process, and many men escape it altogether. About one in five starts to go bald soon after adolescence, although the change is so slight at first that it is hardly noticed. By the age of thirty, however, the balding 20 per cent will have become well aware of what is happening. By the age of fifty, some 60 per cent of white males will have shown some degree of hair loss. The figure is lower for other races.

There are four main, genetically determined, pathways to

advanced baldness. They are: the Widow's Peak, the Monk's Patch, the Domed Forehead and the Naked Crown.

In the Widow's Peak the hairline recedes more and more from the two temple regions, leaving a narrowing strip of hair along the centre line of the head. The Monk's Patch retains the frontal hairline which holds firm but a bald spot starts to grow at the back of the top of the head. The baldness spreads steadily from that region. The whole of the frontal hairline starts to recede in the Domed Forehead, creeping further and further back. Finally, in the Naked Crown the frontal hairline recedes fast in the central region and slower at the sides, the opposite of the Widow's Peak pattern.

To complicate matters, these four main types of balding may be combined, so one individual may display two patterns at once. Genetic factors control these differences, and if a balding man looks at old photographs or paintings of his male ancestors he will usually find that his hair papillae are following in a long family tradition.

Because it is linked with high levels of male sex hormones and because it increases in extent with advancing age it is obvious that baldness is a human display signal indicating male seniority and dominance. It typifies the virile older man and separates him visually from younger, hairier males.

As more time passes, and men become elderly, they experience a waning of their sex drives. Logically, bald heads should then start to sprout hair again, but this does not happen. It seems as if, after years of inactivity, the hair papillae are no longer capable of being revived. They have not been suppressed, they have been destroyed.

At this point the link between powerful male sex hormones and a man's bald head would start to become something of a cheat, but a new signal is then added that converts the senior virile male image into that of a grand old man: his hair turns white. This happens to both bald and to hairy heads and transmits the vital I-am-very-old signal in both cases.

On the subject of pigment loss, we often speak of grey hair, but in reality there is no such thing. Individual hairs do not turn grey; they stop producing pigment and turn white. So-called grey hair is simply a mixture of old hairs that are still their original colour, and new hairs scattered among them, that are pure white. At first the white hairs make no visual impact, but as they increase in proportion to other hairs they give the overall impression of greyness. Eventually, as more and more follicles give up pigment production and create unpigmented hairs, these white strands swamp the few remaining coloured ones and at last replace them altogether, leaving a shock of white hair to signal that old age has arrived.

Before considering social attitudes to hair, there are a few remaining anatomical details to record. Yet another oddity of the human hair pattern is that it lacks feelers or vibrissae, the tactile hairs we know so well in their familiar form as cat's whiskers. All mammals, even whales, have at least a few vibrissae but humans alone are without them. Another missing element is the ability to make our hair stand on end when we are angry. Many mammals are capable of bristling with rage and making themselves appear much larger in the process, but humans have lost this dramatic transformation display. This is not so surprising. Erecting our short, sparse body hairs wouldn't frighten a mouse, and our head hairs are obviously too long and too heavy to be hoisted into the angry-erect condition.

Despite the lack of hair-erection displays humans do, however, still retain the tiny muscles that move the hairs. Called the *arrector pili* muscles, the best they can do for us today is to produce goose pimples when we are frozen with cold or fear. What they are doing is an attempt to thicken our non-existent coat of fur. If we still had our fur, this thickening would increase the insulating layer of entrapped air and in this way would help to keep us warm.

There is, however, an exception to this general rule about the inefficiency of our hair erection. It concerns the reaction to the sound of a creaking door, late at night in a dark house. 'It made my flesh creep' is the usual description, and this creeping sensation is, of course, the thousands of *arrector pili* muscles contracting. Sometimes people say 'My hair stood on end' and they comment that the sensation was strongest on the back of the neck. This is probably because it is there that the hairs are dense and short enough to be able to produce a particularly strong local response.

One way in which human hairs do not differ from those of our mammalian relatives is in the presence of associated sebaceous glands.

These tiny glands, situated at the side of the hair root inside the follicle, produce an oily secretion, the sebum, that helps to lubricate the hairs and keep them in good condition. Overactive sebaceous glands produce greasy hair, underactive ones give rise to dry hair. Hair-washing is important to remove dirt from the hair, but it also removes the natural sebum, and too much washing can be almost as damaging as too little.

The strength of healthy human hair is remarkable. Chinese circus acrobats have been known to perform tricks while suspended by their hair, without undue discomfort. A single strand of Chinese hair is reputed to have a breaking strain of 160 grams. It is also highly elastic, and can be stretched by as much as 20 or 30 per cent before snapping.

Hair colour varies in a rather simple way, largely following the colour of the skin. The same pigmentation system is used in both cases. People living in sunny countries have a large number of elongated melanin granules in the cells of their hairs, making them appear black. People in more temperate zones have slightly less melanin, giving their hair a brownish colour. Up in the cooler, sunless world of Scandinavia, there is even less melanin, resulting in the pale-coloured hair we call blond. Albinos, who lack melanin completely, have pure white hair.

This simple scale of black to white is complicated by one rogue element. Certain individuals have melanin granules which, instead of being granulated, are spherical or oval in shape. These are seen by the eye as red. They may occur by themselves, with none of the usual elongated melanin granules, in which case their owner will appear as a golden-blond. If they exist in combination with a moderate amount of the ordinary granules, then their owner will have a rich, reddish-brown tinge and will be referred to as a fiery redhead. If the spherical granules occur in combination with large numbers of elongated granules, then the blackness of the hair will almost mask the redness, but it will still be present to give a subtle tinge to the hair and make it different from the pure black variety.

The shape of each hair varies considerably, but three main hair types are now recognised: the crinkly (*heliotrichous*), typical of Africans; the wavy (*cynotrichous*), typical of Caucasians; and the straight (*leiotrichous*), typical of Eastern races. It is usually claimed that these three racial hair types are related to climatic conditions, but several objections have been raised against this view. It is accepted that the tightly looped hairs on the African scalp do create a bush-like barrier between the skin and the outside world, where the sun is beating down viciously on top of the human head. This barrier creates a buffer zone of entrapped air that helps to prevent overheating of the skull. But if such a buffer zone is so efficient in hot climates, it is argued that it would be equally valuable in cold climates where the entrapped air would act as an insulating device.

Looking at the wavy hair of the Caucasians, two more problems arise. To start with, Caucasian hair is very variable, ranging from almost straight to curly, without any reference to shifts in the environment. Secondly, Caucasians live successfully in a huge range of habitats, from the frozen northlands of Scandinavia to the searing heat of Arabia and India. A similar problem arises with the straight hair of the Eastern races. It may show little variation, but it has an even wider range from north to south.

The main argument to set against these objections is that recent migrations of human beings have spoiled the original pattern. Suppose, for a moment, that in the beginning there were crinkly-haired people in hot countries, wavy-haired people in temperate countries and straight-haired people in cold countries. The crinkly hair would combat the overhead sun without hanging down to prevent sweating from the neck and shoulder region. The long, straight hair would act like a cape over the neck and shoulders, keeping them warm. And the wavy hair would be a compromise between these two extremes, suitable for the intermediate zone. Then, from this starting point, huge migrations would take place, too swift for genetic alterations in hair type to keep up with them.

This scenario makes sense, but it is little more than guesswork. The fact that the three main types of hair *look* different may also have been playing a role. If at some point in the distant past the three major races of mankind were pulling apart from one another, they might have used visual differences of this kind as isolating mechanisms and the three hair types might have been important as racial 'flags'. This might have kept the three types going even after migrations had taken their owners into inappropriate climatic conditions.

Up to this point hair has been considered in its natural condition but the busy fingers of humanity have rarely left it in that state. The urge for self-decoration has produced some startling modifications and distortions. The most basic and widespread attack has been to interfere with the natural length of the hair.

Hair-length alterations have nearly always been linked to the introduction of a new gender signal. As already explained, male and female hair in the natural state shows little difference in this respect so that it is quite arbitrary as to which sex gets the 'short straw' in any given culture. For some tribal societies, the males enjoy elaborate hairstyles while their females display shaven heads. In other cultures, the long tresses of the females are their crowning glory, while the males clump about in bristly scalp-stubble.

As a result of these hair-length contradictions, two completely different types of hair symbolism are interwoven in human folklore. In one strain, the male's great mop of hair is seen as his strength and his virility, giving him power, masculinity and even holiness. The word Caesar, for instance, and its derivatives Kaiser and Tsar, meant hairy or long-haired and were thought eminently suitable for great leaders. This tradition goes right back to the earliest of hero figures, the Babylonian Gilgamesh, who was both long-haired and immensely strong. When he grew sick and his hair fell out, he had to go on a lengthy journey so that '. . . the hair of his head was restored . . .' and he could return refreshed, with his mighty strength renewed.

In this folklore tradition, the virility of the male's copious head of hair is undoubtedly linked to the fact that, although sex hormones make both male and female bodies hairy at puberty, it is the masculine body that becomes hairier, sprouting not merely pubic and armpit hair, but also beard, moustache and, frequently, straggling hairs on trunk and limbs. If extra hairiness is masculine, then all hair becomes symbolic of masculine power and virility, even the hair on the head.

It followed from this that to shave the head of a male was to humiliate him and that to shave one's own head was a sign of humility. For this reason, many priests and holy men cropped their heads to humble themselves before their deity. Oriental monks went shaven-headed as a symbol of celibacy. Inevitably, psychoanalysts have interpreted the cropping of male hair as displaced castration.

In some religions male hair was worn with pride and was never cut. Sikhs follow this tradition, even today, covering their long hair and keeping it tidy by wearing a traditional turban at all times. This decision to let the hair grow long is taken, they will explain, out of respect for God's creation.

The tradition of long male hair was completely contradicted by no less a person than St Paul, who told the Corinthians that it was natural for a man to have short hair and a woman to have long hair. He seems to have been influenced in this comment by the Roman military custom of cropping soldiers' hair. This appears to have been done, not as a humiliation, but simply as a means of increasing uniformity and discipline and to make the Roman troops look different from their long-haired enemies. An element of hygiene may also have been involved. Whatever the reason, St Paul came to the conclusion that short male hair was a glory to God, while long female hair was a glory to man. For this reason he demanded that men should always pray bare-headed, while women should always cover their heads in prayer, a Christian custom that has lasted for two millennia, despite the fact that it is based on a complete misunderstanding about human hair.

St Paul did not mince his words. At one point he said: 'Does not even nature itself teach you that if a man has long hair, it is a shame unto him? But if a woman has long hair it is a glory to her; for her hair is given to her as a covering.'

Although there have been many attempts to break away from this dictum of St Paul's, its influence is still with us to this day. Despite occasional hirsute rebellions by Cavaliers in the seventeenth century and hippies in the twentieth, the shaggy, long-haired male has remained a rarity, and despite similar rebellions by bobbed and snipped modern females, the short-haired female has also proved to be a rare specimen during the centuries since Paul laid down his ruling.

One possible reason for this may relate to the harshness or softness of the hair. Closely cropped hair is bristly and rough. Long flowing tresses are soft and silky. Rough and soft, male and female, could be an unconscious factor tending to promote an acceptance of artificially shortened male hair. Even in these days of sexual equality we still seem to be unable to regain the natural state of hair-length equality.

Returning for a moment to the question of the balding male, it is important to realise that his condition has little bearing on the long-hair/short-hair debate. A bald man may still present a long-haired or short-haired appearance, depending on whether the surviving ring of hair around his bald patch is allowed to grow and hang down around his head or is closely cropped. But whichever way he wears it, the major hair signal emanating from his head will still be that of a gleaming bald pate.

For centuries this shiny signal has been the cause of male dismay and bald men have often gone to great lengths to conceal their baldness from the eyes of their companions. There is an old saying that sums this up: 'Men who scorn death in battle quail before a receding hairline.'

Bearing in mind that baldness reflects a high level of male sex hormones, this curious obsession requires some explanation. The answer has to do with the strange delay in the development of the bald pate. Although it may start out as a sign of virile masculinity, it takes so long to develop fully that it ends up being more of a signal of advancing age. In a culture that worships youthfulness this is clearly something of a disaster, especially for males who appear in public as actors or singers and who are meant to be trans-mitting sexual charm during their performances. If eighteen-year-old males are at their highest sexual peak (which they are) and if eighteen-year-old males never display bald heads (which they don't), then older males must aspire to copy the eighteen-year-old condi-tion in as many ways as they can. For professional performers reaching middle age this may require a rigid health regime and, above all, the display of a hairy head.

If the hairs desert such males, serious efforts must be made to correct the situation. For them, even the certain knowledge that they are the victims of genetics rather than some curable disease or improper diet is not enough to deter them.

During the course of history, men have taken six main courses of action to combat baldness. The oldest known treatments date from an Egyptian papyrus of 1500 BC. One of the ancient cures mentioned there involved the application of a mixture of fat taken from the bodies of lions, hippos, crocodiles, cats and snakes. Variations on this search for an elixir of youth included removing the prickles from hedgehogs, burning the spikes and mixing them with oil, fingernail scrapings, honey, alabaster and red ochre.

These ancient recipes may seem ridiculous, but the practice of rubbing strange mixtures into the skin of a bald head was to have a long history. Over three thousand years later, in the eighteenth century, a popular European remedy was the oiled ashes of burnt frogs, bees or goat's dung. In the nineteenth century, there was a new recipe: garden snails mashed up with horse leeches, wasps and salt. Clearly, Victorian treatments were little improvement on those of ancient Egypt.

At the beginning of the twentieth century there was a shift of emphasis. Current opinion now held the view that baldness was caused by some kind of infection spread by dirty combs and brushes at barber shops. This neatly explained why women did not go bald – they did not visit the unhygienic barber shops frequented by men. The favoured treatment of the day became a matter of ruthless sterilising and cleansing of the male head, using antiseptic soaps and a whole range of other anti-bacterials.

Many more crank cures were on offer in the twentieth century. As recently as the 1970s it was still possible to read that baldness could be avoided by eating a diet of natural organically grown foods. The author warned, in all seriousness, that if we continued to consume 'synthetically fertilised, artificially prepared foods' we would soon see 'a totally bald human race'. Following the correct diet as a way of avoiding baldness was still being offered as a solution as late as the 1990s, despite the fact that half a century earlier, in 1942 to be precise, the truth, namely that a bald head is due solely to male genetics and hormones and is a natural hereditary condition, had been published by an American scientist.

Such is the male desperation to display a full head of hair that even the most outlandish and eccentric theories managed to outlive this truth and, even today, in the twenty-first century, quack medicines are still on sale and billions are still spent on elixirs and compounds that claim to make the hairs on the bald patches grow and flourish once more.

The older lotions, oils and greases are usually spurned, but every few years some new chemical is invented that seems to offer a ray of hope. In the 1980s, for example, a substance arrived on the scene called minoxidil. This is a blood-vessel dilator, and its hair-growth potential was discovered by accident. It was being used to control blood pressure and in this role was administered orally, in tablet form, to a man who had been bald for eighteen years. Within four weeks of the treatment, normal dark hair was growing on top of his head. The excitement over this amazing side effect was slightly dampened by the fact that he was also sprouting hair on his forehead, his nose, his ears and other parts of his body.

Switching to a locally applied solution containing minoxidil, doctors then started rubbing the bald heads of selected patients and claimed that with certain kinds of baldness they had an 80 per cent success rate, even though the hair growth was rather modest. Sadly, it was later discovered that when the treatment was stopped the hair disappeared again, so the applications had to continue forever if the treatment was to be successful.

As time went on, minoxidil was being widely marketed and it did, indeed, prove itself to be the first scalp application to have any true impact on hair growth. What was disappointing was that it had this impact with only a small percentage of the males who used it. An article in the medical journal *The Lancet* concluded that for over 90 per cent of bald men, the application of minoxidil was useless. Where improvement did occur, the increase in hairs was, at best, only up to 17 per cent of the density of a full head of hair. So, although minoxidil, unlike all the quack, snake-oil remedies of the past, does have a genuine effect on hair growth on bald male heads, it is an effect so limited as to be of little interest to most men.

Such is the desire to look forever eighteen that intensive research in this area is continuing and several more chemicals, such as finasteride, have now been found that appear to have similar effects to minoxidil. None of them gives back precisely what is wanted – a full, youthful head of hair – but they do all promote the sprouting of a few brave hairs where none existed before, and that is enough to see hundreds of millions being spent each year by the more desperate of men.

A second and more drastic action for today's bald-headed males is to undergo a surgical procedure in which hairy tissue is transplanted on to the bald patch on top of the head, a process known as micro-surgical hair-plugging. Early attempts at this were often disastrous, as one famous pop star knows to his cost. More recently the techniques have been improved, as one former Italian Head of State has ably demonstrated.

Far less drastic is the Sidewinder technique. All this involves is a careful combing and brushing of hair from the side of the head so that it spreads over and covers the bald pate. Unfortunately, although it covers the patch of bare skin, it does so in such an unnatural way that it fails to conceal the fact that baldness has arrived.

A fourth method, much favoured by bald actors who enjoy outdoor sports, is the wearing of exotic hats to cover their gleaming shame. The famous crooner Bing Crosby, who loved golf, was never, ever seen on the course with a bare head. He died after a round of golf and one can imagine him lying there, still with his hat firmly in place.

Fifthly, there is that popular old stand-by, the wig, hairpiece or toupee. In ancient Egypt the pharaohs and their families shaved their heads and then covered them with ceremonial wigs. Slaves were forced to wear their own hair, by law. This high-status role for wigs was repeated elsewhere. Wigs are known to have been worn by Assyrians, Persians, Phoenicians, Greeks and Romans. It was in seventeenth and eighteenth-century Europe, however, that stylised wigs reached their zenith. Although they were introduced during this period as a means of concealing baldness, they soon spread to become a fashionable mode of dress for all ranking members of society and were even worn by schoolchildren at top boarding schools. By the 1750s the fashion was dying out and soon disappeared altogether in ordinary society.

Nowadays wigs appear only as secretive, realistic hairpieces. The one exception, in certain countries, is the survival of the long, stylised wig on the head of the judge in the antique atmosphere of the courtroom. Even this is now under threat, with the British Law Society issuing a statement that 'It is important that court users should not feel intimidated or alienated by what they see in court and therefore the Law Society favours abolishing wigs in all cases.' Many voices have been raised against this recommendation and it remains to be seen whether it will be adopted in the future, heralding the final disappearance of the conspicuously stylised wig of old.

Finally, there is the habit of shaving the entire head, so that it is impossible to distinguish a bald patch on top of it. By scraping away the scalp hair that would normally survive baldness, this gives the impression that the men concerned have deliberately chosen to do away with all head hair. In terms of association this puts them into one of several categories – humble Oriental monks, ancient rulers, sheared criminals, or professional wrestlers. Their dominant, active lifestyles narrow this list down, so that they emerge with the personalities of wrestler-kings, tough men who scorn orthodox fashions and who appear dignified, yet ready for a fight. Compared with the sneakiness of the toupee-wearers, there is something brave about their blatant defiance of the laws of hairiness, and they come out easy winners.

Those are the six voluntary methods of combating baldness, but the best way of all is to be born into a family where all male ancestors were fully haired into their dotage. If the baldness genes are completely missing from your family tree you will never need a toupee. And there is, of course, a guaranteed method that will ensure a full head of hair all your life. If you are castrated before you reach puberty, thus removing the main source of testosterone, you will never lose your hair. There were no bald eunuchs in the Sultan's harem.

There is one curious form of artificial baldness, namely the cleric's tonsure. This involved the deliberate shaving away of the hair from the top of the head to create an area of naked skin. Traditionally, this was the hairstyle of the devout monk and symbolised the fact that he had given up all thoughts of worldly fashion and concern with personal appearance. It was meant as a sign that he rejected society's standards, but it did more than that because it also gave the monk a clearly identifiable visual label. Although it was a kind of self-inflicted baldness, it was always easy to distinguish it, at a glance, from true baldness.

In early days there were three types of tonsure. The Oriental involved the shaving of the whole of the head. The Celtic consisted of shaving the whole front of the head from ear to ear and the Roman consisted of shaving only the top of the head, allowing the hair to grow in the form of a crown. This is claimed to have originated with St Peter, and was the style adopted by the Catholic Church until 1972, when obligatory tonsure was abolished. Certain orders of monks, including Carthusians and Trappists, have ignored this abolition and continue to display the tonsure, even today.

Turning to the general question of hair decoration, it is safe to say that there is no society or culture anywhere on the globe today that does not decorate or style the hair in some manner. This has been the case for thousands of years. Hair has been dyed, shaped, lacquered, curled, straightened, powdered, bleached, tinted, waved, braided, dressed, greased and oiled in a million different ways at a cost of countless hours of human labour and ingenuity. One of the reasons for the special attention paid to this part of the human anatomy is that hair can be altered so easily and, above all, because, after it has been cut or modified, it eventually grows back again. This constant renewal of scalp hair has made it into a suitable symbol of the life force itself and has loaded it with a huge variety of superstitious beliefs and taboos.

The giving of a piece of hair in a locket to a loved one was an act of total surrender to him or her, a symbolic placing of your soul in the other's power. The lock of hair contained the vital spirit of the giver and, by wearing it around the neck, the loved one was given the power to control and bewitch the donor. An unusual variant of this concerns the chivalrous knights of the Middle Ages. Dedicated to courtly love, these brave warriors wore a tuft of their mistresses' pubic hair in their hats when they went into battle.

Because of its magical powers, barbers in superstitious societies were forced to bury the shorn hair of their customers in secret places so that it could not be stolen and used in magical ceremonies to harm them. Such customs are by no means dead today. Parents in rural areas of some European countries are still warned not to keep locks of their children's hair if they wish them to live a long life. This is because, once again, it is feared that evil beings may get hold of the clippings and use them to cast a spell on their owners.

People rarely touch one another's scalp hair unless they are lovers, parents, priests or hairdressers. The head region is a protected area and is not available to casual acquaintances, largely because of its close proximity to those precious and extremely delicate organs, the eyes. Only the most highly trusted companions are allowed to make contact with the head hair. When this happens a number of characteristic actions are observed. There is the laying-on of hands, when a priest blesses a believer. There is the pat on the head of a child by a proud parent, or the mocking pat on the head between adult males suggesting condescendingly that one of them is behaving like a little child. There is the intimate hair-to-hair contact of lovers who put their heads together and, later on, the hair-stroking, fondling and kissing of love-making.

Above all, there are the many hours of life spent having the hair groomed by professional hair-touchers, the barbers and hairstylists. This goes far beyond the demands of cleanliness and even beyond the need for decoration and display. It harks back to those distant, primitive times when, like our close relatives the monkeys and apes, we spent lengthy periods of each day grooming one another's fur. As with all primates, this activity was much more than a comfort activity; it was a method of cementing social friendships within the group. It was caring, non-aggressive physical contact with another being and this gave it a deeply rewarding feeling. It is the same pleasant sensation that people feel today, millions of years later, when they give themselves up to the luxury of the hair-groomers' hands.

There are a few exceptional male hairstyles that deserve a special mention. The most bizarre is that of Aaron Studham from Massachusetts, a teenager with a Mohican crest that stands 24 inches (61 cm) tall. It took him six years to grow and, rather like a cockatoo, he can wear it up or down. When he is dressing with it in the upright position it takes him forty-five minutes to prepare it, using copious hairspray.

The Mohawk style has a special appeal to extrovert males and always causes comment because of its extreme shape, with the central hair excessively long and the side hair cropped short. It is based on the style adopted by Amerindian Mohawk tribal warriors, and has the advantage of greatly increasing the height of the wearer.

Although extremely rare, the Mohawk has already developed a cult following and has ramified into several distinct sub-styles. There is the Liberty Spikes Mohawk, cut into tapering points in imitation of the Statue of Liberty. In the Dreadhawk, the central hair is dread-locked instead of being groomed erect. In the Deathhawk, a favourite of Goths, the long central hair is worn wider and looser. The Bihawk has two separate, central crests of long hair. And in the Reverse Mohawk, the central hair is shaved and the lateral hair is grown long.

By tradition, a Spanish matador must wear his hair in a pigtail. When he finally retires from the bullring, he is said to be cutting the pigtail. The origin of this type of pigtail goes back to ancient Rome, where it was the caste mark of gladiators who fought bulls in the Colosseum. In the eighteenth century, Spanish bullfighters wore their long hair in a net to keep it out of their eyes. Later they tied their hair in a knot and then, eventually, adopted the pigtail, which became a badge of their profession.

The famous Chinese pigtail, or queue, in which the whole of the skull was shaved except for the long tail at the back of the head, was introduced in the seventeenth century by the semi-nomadic Manchu people from Mongolia in the north who issued a Queue Order requiring that this hairstyle be worn by all Chinese men. Failure to do so was punishable by death and this led to numerous rebellions, with tens of thousands of Chinese men dying because of an imposed hairstyle. The reason for their fanatical opposition was that their own tradition stated that the removal of hair was against filial piety, because a man's hair was a gift from his parents. They lost their struggle and the pigtail style became the rule until the early part of the twentieth century.

In the nineteenth century Chinese immigrants in San Francisco were being persecuted by the local law-makers, who insisted that any Chinese men sent to jail should have their pigtails cut off. By this date, pigtails were worn with pride, and one Chinese prisoner sued when his was forcibly removed, alleging that the loss of his queue '. . . had exposed him to public contempt and ridicule and had irreparably injured him in the eyes of his countrymen'. He won his case and pigtails were never attacked again. In 1911, when the Manchu Dynasty fell, all Chinese males adopted a modern style and then willingly removed their traditional pigtails.

Another extreme form of male hairstyle is the dreadlock. Dreadlocks hang like thick cords of hair all over the head. They are matted ropes of hair that form by themselves if the hair is permitted to grow without any combing, brushing or washing over a period of years. They have appeared in cultures all over the world from ancient Egypt, ancient Asia, and ancient Mexico, to the Celts and the Vikings. Today they are best known in the form of the Jamaican Rastafarian Rastalocks. The Rastas started wearing their hair in this manner early in the twentieth century and claimed that they were following in the footsteps of important dreadlocked figures such as John the Baptist and Samson. They took their stance from the Bible, quoting the Book of Numbers: 'There shall no razor come upon his head: until the days be fulfilled . . . he shall be holy, and shall let the locks of the hair of his head grow.'

Today, dreadlocks are more a symbolic form of rebellion against the Establishment, specifically against a Eurocentric Establishment. In the 1980s, Bob Marley and reggae music made this type of hairstyle so popular that it soon spread into the very culture it was opposing. Fashion trend-setters began dressing their more extrovert clients in dreadlocks and before long it was possible to have a whole range of dreadlock extensions. These were artificial dreadlocks that could be fitted to the real hair in a few hours, instead of having to wait for the real thing to grow, over a period of several years.

Hand gestures involving the hair are few. There is the Hair Clasp, in which the hand is brought rapidly up to clasp the top of the head. This is done unconsciously when a man has suddenly realised that he has done something stupid. It is a self-clasping action that acts as a self-comforter. It carries the message: at this moment I need to be clasped protectively, as I was when I was a child, but now as an adult I must clasp myself. If the moment of stupidity is an acute one, as when a footballer misses an open goal, there may be a double-strength action, with both hands coming up together to clasp the top of the head.

There is also the common Head Scratch of the puzzled man. In this case, it seems that the conflict he is experiencing is disturbing the secretions of his skin glands and this is making his scalp itch, triggering the brief, scratching action.

There is one form of head scratching that has a special meaning, and that has to be separated from the puzzled action. This is the scratching of the *back* of the head. It is done in moments of frustrated aggression and is derived from our primeval attack movement. When angered and about to strike someone, we automatically raise the arm to deliver a downward blow. The frontal blow of the trained boxer is a much more sophisticated movement and has to be learnt, but even tiny children doing battle in the nursery employ the overarm blow, and it stays with them throughout their lives. If as adults they become involved in street rioting they will revert to it again, and the riot police will respond in a similar manner, beating them over the head with clubs. A man in an angry but inhibited mood at a social gathering cannot thump the person who is annoying him, but his arm flies up as if to do so, responding to the primeval prompting of his unconscious mind. When the arm reaches the uppermost position, ready to start on its downward arc, it is checked and the impotent hand diverts itself with a vigorous scratch or pat, as if to suggest that that was the intention all along.

3. The Brow

The human brow, made up of forehead, temples and eyebrows, is the direct result of our ancestors' dramatic brain enlargement. The brain of a chimpanzee has a volume of roughly 400 ccs; for modern man the figure is 1,350 ccs, more than three times that of our hairy relatives. It was the expansion of the human brain, especially in the frontal region, that gave us a 'face above our eyes'.

If you look at a chimpanzee's face side by side with a human face, the forehead difference is striking. In the case of the ape it is almost non-existent. In the human it rises vertically above the eyes as a great naked patch of skin. The chimpanzee's hairline comes right down to the eyebrows, which are almost hairless. In fact, the brow region of the ape is the complete opposite of that in human beings.

The difference in brow-ridges also demands explanation. In apes they are bony crests over the eyes that protect them from physical damage. The early ancestors of man also possessed these heavy rims of bone, but they gradually dwindled in size until today they are almost gone. Why did we lose them? When we became primeval hunters we would surely have needed them even more than in our remote fruit-picking days.

The answer is that their loss is more apparent than real. If profiles of the head of an ape and of a man are compared it emerges that the protective line of the brow-ridge remains roughly where it is, while the forehead expands above it. By the time the condition of modern man has been reached, the forehead, inflated by the ballooning human brain, has pushed forward to the level of protrusion of the ancient brow-ridge. So the new forehead acts in the same protective way, providing a bony defence against blows to the eyes. The brow-ridges have not vanished, they have simply been engulfed.

It could be argued that, with the increased hazards of violent hunting activities, it might have been an advantage to have doubled the protection, to have kept the bony ridges in addition to the bulging forehead. One suggestion as to why this did not happen is that during the Ice Age our shivering forebears developed more flattened faces as a defence against the cold. The move towards flattened, fat-lined faces of the kind we still see today on Eskimos included a reduction of the sinuses over the eyes, sinuses that became vulnerable to infection in the colder climate. This reduction had the effect of flattening the brow.

The difference between the eyebrows of ape and man is also intriguing. In both cases the main evolutionary effort appears to have been to make them conspicuous and contrasting with their surroundings. A young chimpanzee has pale, naked eyebrows that stand out vividly against the dark hair above them. A young human has dark eyebrows that stand out against the pale skin above. Even in dark-skinned races the contrast is not lost, so that movements of the eyebrows are still clearly visible to those nearby.

It has often been said that the eyebrows exist as deflectors, helping to keep liquid from trickling down into our eyes when we are sweating or out in a heavy rainstorm, but anyone watching an exhausted sportsman wiping dripping sweat off his forehead will know that, in this role at least, they are of little help. There is little doubt that the primary function of these conspicuous superciliary patches, as the eyebrows are known technically, is in reality to signal the changing moods of their owners.

The eyebrows of the human male are heavier and bushier than those of the female and contain more individual hairs, suggesting that it is the males, perhaps, who have the greater need to express their mood shifts clearly. There are four expressive muscles that control the position of the eyebrows. The *frontalis* elevates the eyebrow, creating horizontal creases in the brow skin. The *Orbicularis oculi* closes the eyelids and pulls the eyebrows towards the eyes. The *Corrugator superiocili* squeezes the eyebrow inwards and downwards at their inner end, creating vertical furrows between the eyes. The *Proceros* lowers the eyebrows.

Using these muscles in various combinations it is possible to make the following facial expressions:

Lowered Eyebrows – the Frowning Brow. The expression of the angry man. His anger makes him defensive and he lowers his brow to protect his eyes from the retaliation that he expects his anger to provoke.

Raised Eyebrows – the Furrowed Brow. The expression of the surprised or frightened man. By raising the brow-skin, he widens his field of vision, increasing his awareness of whatever it was that caused his fear.

Cocked Eyebrow – the Quizzical Brow. The contradictory expression of a man who is sceptical, part-angry and part-fearful, with one eyebrow lowered and one raised. For some reason, it is more commonly performed by adult males than by females or children.

Oblique Eyebrows – the Knitted Brow. The expression of grief or acute anxiety, with the eyebrows pulled tightly towards one another and raised at their inner ends. It is the expression of chronic pain, or of the mourner.

Flicked Eyebrows – the Greeting Brow. The eyebrow-flash expression of a man who has just seen a friend and is acknowledging their presence. It is a worldwide greeting signal of our species, in which the eyebrows quickly flick up and then down again.

Bobbing Eyebrows – the Mock-erotic Brow. A joke expression. If the flicked eyebrows say 'Hallo', the bobbing eyebrows, with the brow-skin rising and falling quickly several times, says 'Hallo, Hallo, Hallo!' It was first made famous by Groucho Marx.

Shrugged Eyebrows – the Discrediting Brow. The expression of the sarcastic complainer. It is the 'I told you so' gesture in which the eyebrows are raised, held in the raised position for a moment, and then lowered again.

The most impressive gender difference with these expressions is found with the frown. The heavier brow of the male face and the thicker, bushier eyebrows mean that, when angry, the male looks distinctly more threatening than the female. The glowering face of the aggressive man is a valuable support system for his threatening vocalisations, whether these be verbal insults or non-verbal grunts and snarls.

A man by the name of Frank Ames is proudly proclaimed to be the owner of the longest eyebrow hair in the world, hair that measures no less than 7.6 cm in length, or 3 inches. This masculine feat is solemnly recorded in *The Guinness Book of World Records*, and it is one that no woman would ever aspire to.

Traditionally, throughout history, men have been far less inclined than women to modify their eyebrows or pay any sort of cosmetic attention to them. Because the female eyebrows are smaller and thinner than those of the male, there has been a great deal of cosmetic improvement to make them even smaller and therefore super-feminine. For the male, a similar exaggeration, to create a super-masculine effect, would involve making his larger eyebrows even bigger, but short of hair transplants this is difficult to achieve. The result has been that, for the vast majority of men, the eyebrows have been left in a natural state. If they grow bushier with old age, so be it. The older man lets them bristle out from his forehead as wildly as nature intended, the unmistakable sign of a mature male. No fancy plucking or shaping for him. But there are some exceptions to this rule, especially among the young adult males of today, and they deserve a passing mention.

Some men have felt that bushy or straggly eyebrows are too un-disciplined and have resorted to snipping and trimming to sharpen the outlines. There is no attempt here to alter the size of the eyebrows, merely to make them look a little neater. This is a simple enough procedure, requiring nothing more than a pair of small scissors and a mirror, but inevitably it has been exploited by commercial interests in an effort to make it sound as difficult and complicated as possible.

Certain salons now offer designer eyebrows under the slogan: 'If eyes are the windows to the soul, eyebrows are the frame.' If a man wishes, he can have his eyebrows reshaped by an eyebrow designer, using wax and some judicious tweezing, but if he feels the urge to get framed in this way it can cost him $100 to acquire what the salon calls a 'manly arch'.

Nor does it stop there. Youth fashion today includes a whole new range of eyebrow mutilations. A popular one is the Beat-up Boxer look, in which there appears to be a vertical scar on one eyebrow that has healed in such a way that no hairs grow there, leaving a narrow gap or slit. This is a style first popularised in 1954 by Marlon Brando in *On the Waterfront*. Recently it has been favoured by trendy males in an exaggerated form, with one, two or even three narrow, vertical, razor-cut stripes at the outer or temple end of one eyebrow. Asked why this style has become popular, one teenager explained: 'So you could look nice and stylish and it seemed like you were in a knife fight on the street.'

This may explain why such a look is popular with certain youth gangs in North America. Hispanic gang members who belong to the SUR 13 gang in California, for example, sometimes leave a single gap, or stripe, on one eyebrow and three gaps on the other one, to signify the number 13. And the northern Virginia gang MS-13 shave their eyebrows in the same way. Amusingly, their display is back to front because they shave one stripe in the left eyebrow and three in the right. This looks correct when they make the razor cuts in the mirror, where they see I on the left side of their face and III on the right side, but when they meet another gang member, what their friend sees is III and I, reading 31 instead of 13. Some youth gangs, it seems, have yet to fathom the intricacies of mirror imaging.

Because these gang fashions are seen to be cool, they have now been taken up by members of the jet-set celebrity circuit. The Hilton Hotel heiress, Paris Hilton, was recently photographed with boyfriend Paris Latsis, said at the time to be the fifty-fourth richest man in the world, who was sporting a shaved middle part of his right eyebrow.

Instead of eyebrow stripes, some younger men indulge in more painful brow-piercing, with the insertion of a decorative silver ring in the fleshy ridge of the eyebrow. This modification is usually placed towards the outer end of one eyebrow. Some mutilatees prefer an eyebrow stud, rather than a complete ring.

For those men who fear that their furrowed brows are revealing telltale signs of ageing, there is help at hand today in the form of a brow-lift facial rejuvenation procedure. Most men in their fifties start to show permanent frown lines and forehead creases that refuse to disappear even during moments of calm and serenity. These lines, caused by a combination of loss of elasticity, exposure to the sun and repeated frowning, squinting or eyebrow raising, can be eliminated by a tight restretching of the brow-skin. Most men who seek this treatment do so because friends have asked them why they are sad, angry or tired, when in reality they are none of these. When they discover that their brow-skin has become stuck in positions that suggest these moods, they sometimes decide to take drastic action and resort to cosmetic surgery.

Some unusual males possess a monobrow – the left and right eyebrow being joined across the top of the nose to create a continuous hairy line. To most men this appears to be too animalistic – a sort of wolf-brow or vampire-brow – but others relish their extra patch of hair and even cultivate it. There is a website called *monobrow.com* where such individuals exchange views. There, you can, if you so wish, join their number by acquiring a monobrow toupee '. . . an adhesive prosthesis designed to match your own eyebrow type and which can be changed quickly and easily'. If you attach this to your face at the top of your nose, it is claimed that it will equip you with the enviable ability to prevent sweat from trickling down your nose and coming to rest in a large blob hanging from your nose-tip.

Because the monobrow creates a dark patch at the top of the nose, it tends to give the impression that a man is frowning, even when he is not. So it makes him look fierce as well as unduly hairy. For this reason many women find it an unattractive feature and some monobrowed men pluck or shave off the offending hairs, with varying degrees of success. Others, however, prefer to remain as nature intended and wear their monobrows, sometimes called unibrows, with pride. Some famous names included in this defiant group are Russian cosmonaut Salizhan Sharipov; actors Colin Farrell and Josh Hartnett; musicians Chris De Burgh and Liam Gallagher; Soviet leader Leonid Brezhnev; Deputy-Führer of Nazi Germany Rudolf Hess, British politican Denis Healey; footballers Eric Cantona, Ronaldo and Wayne Rooney; and tennis champion Pete Sampras.

The general height of the brow region has given us several popular terms: highbrow, middlebrow, lowbrow and no-brow. The term highbrow dates back to the middle of the nineteenth century, and grew out of the popularity of phrenology. It was thought that an unusually tall forehead indicated greater intellectual power and the word was originally used as a compliment. But as time passed, its meaning shifted so that it was employed as a mild insult indicating that someone was supercilious and snobbish. The term lowbrow was introduced as the opposite, indicating something vulgar and coarse. Then, in the 1940s, *Life* magazine coined the term middle-brow to indicate someone whose tastes were neither high nor low, but moderate and conventional. More recently a staff writer on the *New Yorker* magazine introduced the concept of the no-brow – someone who simply does not fit into the class structure of high-, middle- and lowbrow – claiming, 'Today's popular culture is much more eclectic than the old split between highbrow and lowbrow'. Unfortunately, this last term, although meant to indicate someone who is outside the class system, conjures up the image of a primi-tive, browless, ape-headed man.

Finally, returning to the body language of the brow region, in addition to the seven positional shifts of the skin and eyebrows mentioned earlier, there are also several hand-to-brow gestures that transmit signals from this part of the body. Some are local and some global.

These hand gestures are always used much more often by men than by women. They include several different versions of the 'You are crazy!' insult. There is the Neapolitan brow-tapping gesture, in which the thumb-tip and forefinger-tip of one hand are squeezed together, as if holding something very small. They are then tapped against the glabellum, the region of the brow at the top of the nose, between the eyes. The message of this gesture is 'Your brain is so small I could hold it like this between my thumb and forefinger.'

There is also the temple-circling gesture, in which the stiff forefinger is rotated at the side of the forehead, meaning, 'You are crazy, your brain is rolling round and round' or, alternatively, 'You are making my brain roll round and round'. This is a common gesture in the West, but in Japan it has a special refinement. There, if the finger rotates anti-clockwise it has the same meaning: Crazy! But if it rotates clockwise, the meaning changes and it signifies that someone is vain. At least, this used to be the case, but modern Western influences are now so widespread among the younger Japanese that the distinction is often lost and both kinds of temple circling are now likely to mean Crazy!, as they do in the West.

There is a variant of the temple-circling action in which the stiff forefinger is twisted as if it is a screwdriver tightening a loose screw. This means either 'You are crazy!' or 'I am crazy!' in much of the Western world. And shooting the temple sees the stiff forefinger placed against the side of the forehead as if it is the barrel of a gun. This is a self-insult saying 'I am so crazy I should blow my brains out'.

Tapping the temple, in which the forefinger is touched several times against the side of the forehead, confusingly has a double meaning. It can either signify that someone is crazy or that they are clever. As an action, it simply points at the brain, suggesting either that the brain requires attention or that it is working unusually well.

There are three animal gestures, in which the hands are brought up to the sides of the forehead to mimic either the horns of a bull, the antlers of a deer or the big ears of a donkey. The first two are Italian cuckold gestures, suggesting that you are so pathetic that your wife is unfaithful to you, and the donkey ears gesture is a Syrian sign meaning 'You are an ass!'

There is also the forehead clasp, in which the palm of the hand covers the forehead and shields the face, a worldwide gesture of depression, defeat and despair. In ancient Greece there was a formalised version of this in which the fist was repeatedly struck again the centre of the forehead. In modern times, we often slap our hand against the forehead in an 'Oh, no!' reaction, when we suddenly realise that we have done something stupid.

4. THE EARS

Compared with other expert hunters, like wolves and lions, the human male has rather modest ears. He cannot prick them up to catch a tiny sound in the distance, nor can he twist and turn them to pinpoint the location of a noise. He is unable to flatten them against his head to protect them when he is fighting, although to be fair they are pretty flat already. He compensates for these shortcomings by having a much more mobile neck than his ancient rivals. Instead of twisting his flexible ears, he twists his flexible neck, and using this method he is capable of detecting the source of a sound to within 3 degrees.

Despite their modest size and their immobility, the value of our external human ears should not be underestimated. Anyone who has had the misfortune to lose his ears (ear amputation was once a recognised form of punishment in England) will know that without them the sounds we hear are much more distorted. The strange arrangement of folds and ridges that we take so much for granted are, in fact, a subtle sound-balancing system that we use every day without giving it a passing thought.

Each ear has a narrow, folded rim called the *helix*, inside which there are various bumps and folds including the *tragus*, *concha* and *scapha*. These surround the opening to an inch-long ear canal down which sound waves pass to strike the eardrum. This canal is lined with four thousand wax-producing glands that produce a yellow secretion that helps to repel insects.

There are marked racial differences in ear-wax production. Members of black and white races nearly all produce a sticky wax, but some Caucasians and all Orientals have a dry wax. Bearing in mind that the sticky wax is a better insect repellent, it is odd that the Chinese, Japanese and other Far Eastern peoples should require less protection in this respect. This is just one of many small oddities of human evolution for which, as yet, we have no explanation.

Two other mysteries are why black people have better hearing than whites, and why women have better hearing than men. When five thousand people were given careful hearing tests, between 1999 and 2004, these racial and gender differences were inescapable. The only suggestions put forward so far are that extra pigment in the skin somehow protects the ears of black people from damage, and that 'melanin plays a role in how the body removes harmful chemical compounds caused by damage to the sensitive hair cells in the inner ear'. The reason given for women hearing better than men is that men are exposed to more loud noises than women during their lives, thus weakening their sensitivity to sound. Neither of these explanations is wholly convincing, but at present there are no other suggestions on offer.

There are also protective hairs at the entrances to the ear canals, and these grow much longer in men than in women. The greatest length recorded for ear hairs is a staggering 4.5 inches (11.5 cm), a record presently held by a teacher in India. He is very proud of his protruding tufts of hairs, but most men prefer to remove them, snipping them off or plucking them.

At the bottom of each ear hangs a unique human feature, the fleshy, rounded earlobe, something that other primates lack. Lacking any cartilage, it does not have any role in the sound-balancing system of the folds and ridges of the external ear, but instead is a soft, smoothly bulbous lump of fatty tissue that appears to function purely as an additional erogenous zone. During sexual arousal the lobes become engorged with blood and highly sensitive to the touch. Sucking, licking, nibbling and kissing them during foreplay can act as a powerful erotic stimulus and this seems to be their only reason for existing.

There are two types of earlobe, the free-hanging and the attached. The free ones are twice as common as the attached ones. This is because the gene for free lobes is dominant and the gene for attached lobes is recessive. What this means is that, if both your parents have free lobes, you will also have free lobes and if one of your parents has free lobes and one has attached lobes, you will also have free lobes. You will only have attached lobes if *both* your parents have them.

This genetic earlobe difference means that, if a woman with attached earlobes is married to a husband with free earlobes and she gives birth to a child with attached earlobes, her husband cannot be the father of that child. Evidence of this kind can be useful in disputed paternity cases.

Turning to mythology, there was once a curious belief that the boneless fleshy earlobe had some connection with the boneless fleshy human penis. Tutors of royal children in the Orient, who were not permitted to punish young princes in a more direct physical way, were allowed to pull their ears when they misbehaved, because it was thought that, by so doing, they would not merely be chastising them, but would also be helping to elongate their penises and thus provide them with increased sexual vigour.

It was another strange superstition that led to the ancient custom of having the ears pierced for earrings. In earlier times people lived in fear of evil spirits, demonic forces that were always attempting to enter the human body through any orifice they could find. The tunnel of the ear was thought to be especially vulnerable and so it had to be protected. It could not be blocked off because this would impair hearing, so the next best thing was to place something precious, some sort of small treasure in the form of gold or silver, as close to the ear as possible. The idea was that, as the evil one was sneaking up on the ear, to bore its way down the inviting ear canal and into its victim's head, it would spot the gleaming metal and would either be distracted by its beauty, or possibly repelled by its metallic magic. In this way the first earrings were used, not as mere decorations, but as life-saving lucky charms.

We know that ear-piercing is at least five thousand years old, because a mummified body of that age found frozen in an Austrian glacier in 1991 had large holes drilled in its ears, and it seems likely that this was the oldest form of body-piercing employed by human beings.

Heavy earrings that pulled the earlobes down and made them longer were especially favoured in ancient times because it was thought that to have very long ears made a man wise and compassionate. A study of early Asian and Oriental sculpture reveals that whenever important persons were being represented, they were always shown with elongated, pendulous earlobes. It was reported that Buddha possessed particularly generous lobes, in keeping with his greatness, because this enabled him to hear the sound of the world and the cries of suffering beings, and to respond to them.

Stretched male ears have also been favoured in some of the more remote hunter/gatherer tribes. In fact, in some Brazilian Indian tribes, decorated earplugs have been almost the only form of adornment worn by the young male hunters. There are early photographs showing them standing stark naked, but with carefully decorated discs inserted into their ears. Anthropologists discovered that they considered their perforated earlobes to be one of the most important of their tribal badges. Among the Timbira Indians of north-eastern Brazil, the ear-piercing operation is part of the initiation cycle of all teenage boys. The perforator uses a wooden pin that he dips into a special pigment to mark the spot on the earlobe where he will insert it. Then, holding a bamboo plug between his lips, he swiftly twists the pin and drives it through the boy's earlobe, making a pencil-sized hole into which he then immediately thrusts the bamboo plug. He then repeats the process with the other ear. It is a mark of pride among the boys being skewered in this way to show no reaction, to make no sound and make no movement.

Once the boy's ears have healed, he will then gradually enlarge the size of the pierced holes by inserting bigger and bigger plugs, until they are large enough to get two index fingers into them. The largest plugs inserted are about 4 inches (10 cm) across and when this stage has been reached, the earlobes are now just a thin strip surrounding what has become a flat, decorated, wooden disc. This form of ear decoration gives the young men of the tribe great sex appeal and it is said that 'such discs are the youth's pride and the women's delight'. The bigger a man's ear disc, the more handsome he is said to be. As he grows older, his disc will only be worn for festive occasions. This leaves him with a problem of what to do with the loosely flapping earlobe rim. His solution is to sling it over the upper edge of the ear and keep it there until he needs it. Alternatively, he can use it in a practical way to carry small objects since, being naked, he has nowhere else to put them.

In Europe, in Elizabethan times, it became popular for sailors, especially pirates, to wear a thick band of gold in one ear. Several reasons have been proposed for this custom. One suggestion is that the sailors believed that an earring would somehow protect them from drowning. Another is that they thought it would improve their eyesight. Yet another is that it would prevent seasickness. It is hard to see how such superstitions could arise or persist, but persist they did to the point where, in the popular imagination, sailors and male earrings became almost synonymous.

A more scholarly theory offers a completely different interpretation, namely that it had to do with the value of the gold from which the earring was made. The idea was that, when sailors set out on a long voyage in an old sailing ship, there was a fair chance that they might never see home again and might die in some foreign land. If this happened it was important to be carrying enough gold to pay for a funeral, but hiding a piece of gold for this purpose was a risky business, so the best place to keep it was where it could not easily be stolen, pressed through a hole in the ear. A variant of this theory suggested that the earring was worn to pay the ship's cooper to make a barrel to carry home the pirate's preserved body, so that he would not have to be buried at sea.

None of these theories seems to have any solid backing from naval records. A more likely, if less colourful, explanation is that sailors originally wore an earring simply because at the time it was fashionable to do so. In the days of Elizabeth I, male earrings were the height of fashion. William Shakespeare wore a gold earring in his left ear and Sir Walter Raleigh wore a large pearl in his. Later on, when the custom had died away on land, sailors seem to have kept it going because by then, in their separate world, it had become something of a naval tradition.

Moving to modern times, in the Western world, earrings, so long a purely female adornment, have recently been seen on increasing numbers of male ears. At first it was assumed that the wearers were all effeminate homosexuals, but it soon became clear that the habit was spreading to the more avant-garde of the young heterosexuals. This led to some confusion and stories began to circulate that there was a secret code, that to wear an earring in a pierced left ear was homosexual, and in a pierced right ear was rebel heterosexual. The problem was that nobody could remember which was supposed to be which. In the end the male earring lost its sexual significance altogether and simply became a generalised way of annoying middle-aged, latter-day puritans.

During the brief flowering of punk rock in the 1970s, the outrage factor was magnified by the bizarre nature of the objects thrust into the clumsily drilled earlobes. Large safety pins were clear favourites, but chains carrying everything from razor blades to electric light bulbs were also used by the shock troops of the new wave.

In the 1980s the male earring spread its range even further, despite mutterings from the corridors of high fashion. Even top footballers were observed signing lucrative new contracts with a fancy diamond stud glinting from one macho earlobe. In adopting this fashion, they were reasserting the young warrior's right to display body ornaments as contrived as any female's. And this trend has survived right through into the twenty-first century, with both males and females displaying ear adornments, although it has to be admitted that female ear-piercing is still far more common than that of the male.

Acceptance of male earrings still tends to be limited to those worn by the younger, more flamboyant males, largely from the world of sport, music and show business. When older, more established males start wearing them it can cause raised eyebrows. In the United States, the greatly respected elder statesman of TV news, Ed Bradley, America's equivalent of the UK's Sir Trevor McDonald, shocked many of his fans when he suddenly decided to insert a glittering diamond-stud earring, a gift from Liza Minnelli, in his left ear and appear in his *60 Minutes* newscasts wearing it. He had his supporters, one of whom remarked: 'Good for Ed Bradley. He even bucked the system with the earring and grey hair and full grey beard, while Mike Wallace has been coloring his hair as old as he is.' But the majority of the viewers who watched the programmes were horrified, saying it made him look pathetic. One said, 'Somebody please tell Ed he's a news reporter not a pirate or a gypsy.' More seriously: 'When Ed Bradley started to wear earrings on the show. The whole news credibility factor fell . . .' In other words, heterosexual male earrings may now be accepted for the younger generation, but not on older men who wish to be taken seriously. That will take much longer and perhaps by then ear fashions will have changed yet again.

Although rare, multi-piercing is also seen today among young social rebels. Instead of one hole, the ear is drilled again and again, all around its perimeter, so that a whole series of earrings can be attached to it. This is a device well known from certain tribal cultures, but previously unrecorded in Western urban societies. In specialist circles today it is known as cartilage piercing and there is a wide range of types available, depending on which part of the ear is mutilated. These include: Tragus piercing, Anti-tragus piercing, Rook piercing, Helix piercing, Orbital piercing, Industrial piercing, Daith piercing, and Conch piercing.

Most of these are self-explanatory, but two of them require some explanation. Daith piercing is comparatively new, having been introduced as recently as 1992. It is done through the innermost ridge of cartilage, right next to the opening of the ear canal, and it must make it appear that the ring inserted into it comes out of the canal itself. It was inspired by the mystical idea that: 'Rings left in an orifice of the body act as a *Guardian of the Gate.*' The superstitious belief is that, from the moment of the piercing, they start to act as a filter. In the case of the ear, this means that they will keep out all that is nonsense and let pass all that is intelligent.

The curiously named Industrial piercing appeared at the same time as the Daith. It has also been called the Crossbow or Scaffold piercing, and it consists of two piercings connected by a single ornament, such as a barbell or rod, that usually runs across the ear, from one side to the other.

The repertoire of gestures and actions involving the ears is severely limited. We cover our ears to reduce noise and we cup them to increase sound. We rub them or tug at them when we are indecisive and cannot think what to say; and when we are alone we often probe them with the little finger, appropriately called the ear finger in some countries, in abortive attempts to clean them.

The most interesting ear gesture is the simple earlobe touch. This has many meanings in different countries. Sometimes the gesturer holds the lobe lightly between thumb and forefinger, sometimes he tugs at it or flicks it with his forefinger. In certain countries, such as Italy and Yugoslavia, to make any of these actions towards a man is extremely dangerous because it implies that he is effeminate, and should be wearing earrings. In Portugal the message is very different. There it indicates that something is particularly good or delicious and may be used to describe everything from girls to food. Clearly, Italians in Portugal might find themselves confused by the reactions to their insult ear-touch, while Portuguese in Italy might be surprised to find themselves in hospital after employing their ear-touch of praise.

A Spaniard would have another interpretation altogether. To him the ear-touch would mean that someone was a sponger, a nuisance who hovers around in bars cadging drinks but never pays for his own. He leaves his friends hanging like an ear lobe. In Greece and Turkey, the ear-touch usually means that the performer of the gesture will pull your own ear if you are not careful. It is a warning to children of punishment to come. In Malta it signals that someone is an informer, that he is 'all ears', so watch what you are saying. In Scotland, it is a gesture that registers disbelief: 'I can't believe my ears.' This is a multi-message gesture, employing a variety of forms of ear symbolism, each association growing up in certain regions and remaining virtually unknown elsewhere.

There is also a dangerous ear gesture that could cause great offence in certain parts of the Middle East, especially Syria, Saudi Arabia and the Lebanon. Called the Ear Fan it consists of placing the tips of the little fingers in the ears, with the other digits spread out like a fan on either side of the head.

This is an Arabian version of the cuckold sign. It implies that the wife of the person to whom it is directed is unfaithful. In origin, it suggests that the victim of the insult should be wearing antlers, like a stag. This contrasts with the more common cuckold sign found around the Mediterranean, where the gesturer makes a horn sign, imitating a bull. In both cases, the action implies that someone is rutting, like a bull or a stag, with the victim's wife. In the strict social world of the Arabs, this is one of the worst insults that can be thrown by one man at another. In some contexts its message would be so potent that it could easily lead to a killing.

Finally, although it is no more than a medical condition, the infamous cauliflower ear deserves a brief mention. It is seldom seen today because treatment for it has improved, but in earlier times it was worn almost as a badge of honour by certain boxers, rugby players and wrestlers. It was caused by the formation of a blood clot inside the external ear when it was repeatedly struck or torn by opponents. If this clot was not given expert treatment, the cartilage inside the ear became separated and then died, leaving the ear permanently swollen and deformed, its strange shape resembling that of a cauliflower.

In ancient Greece, the sport of boxing was so brutal that even the sculptures of famous boxers at Olympia show mutilated ears, and Plato described boxers as folk-with-battered-ears. During training, the Olympic boxers wore special ear-guards called *amphotides*, consisting of two circular pieces of thick leather or metal. These were fastened over the head and under the chin. During the games themselves, however, this form of protection was not permitted and the ears suffered as a result, even though there were strict rules preventing the leather thongs, worn on the boxers' fists, from being reinforced in any way. One of the boxers was even given the name of *Otothladias*, or Cauliflower Ears.

As the years passed, boxing contests in the ancient world became more and more savage, with the fighters' leather-bound fists now wrapped in metal rings, or with sharp metal points sticking out. It is not difficult to imagine the kind of damage that could be done to delicate ears with this hand armour and, as the metal devices grew in ferocity, the fighters defended themselves by wearing helmets with special ear-protectors hanging down from each side.

5. THE EYES

The eye is the most extraordinary organ in the human body. No bigger than a table-tennis ball, it is capable of responding to one and a half million simultaneous messages, and provides us with 80 per cent of our information about the external world, four times the information we receive from all our other senses put together. The male eye is very slightly larger than that of the female, as would be expected with the heavier male head, but there is a great deal of individual variation in this respect.

In the world of caricature, the typically masculine eye expression is the narrowed gaze of the hunter, which contrasts with the attractively wide-eyed look of the female, a look that she may deliberately exaggerate with cosmetics. When training students, a cartoonist will inform them that most male eyes are thinner and narrower than female eyes, and this is a difference that is used to create extreme images of masculinity or femininity. An eye surgeon will tell you that this is nonsense and that male and female eyes show little or no differences, but this is because, as a surgeon, he is concerned with the eyeball itself, while the cartoonist is concerned with the typical degree of opening of the eyelids.

In older males, the upper eyelids often become heavier and start to give the eyes a tired, hooded look. Sometimes this becomes so extreme that cosmetic surgery is called for and a male-eyelid-lift operation is performed. Interestingly, surgeons warn against taking this too far, cautioning that 'One has to be careful with male eyelid lift surgery in order to avoid a commonly seen mistake – creating a feminized appearance.'

This difference in eye-opening underlines the fact that the female eye usually shows more white than that of the male, and it has been suggested that this difference means that men have better forward or tunnel vision, while females have better lateral vision. This would make sense in evolutionary terms, since it would give the male hunter a style of seeing that would help him to focus single-mindedly on distant prey, while the multi-tasking female would be more aware of everything around her. Again, however, there is a great deal of individual variation, and eye surgeons reject the idea that there is any gender difference in the eyeball's mode of seeing. If any difference does exist, it must be the result of the way the male and female eyelids operate, with the female lids normally pulled back more, giving that wide-eyed look.

Having visible whites to the eyes is a uniquely human feature. The 'whites' of a chimpanzee's eyes, for example, are in fact brown. This means that it is much more difficult to see shifts in gaze direction in chimps. When humans are clustered together in a group, however, it is obvious who is looking where, because the varying exposure of the whites tells us exactly how the eyes are changing position from moment to moment. As a result of this we are, unconsciously, able to read one another's eye signals every time we find ourselves in a social situation. We use this information to keep ourselves informed about the relationships that exist between our companions, about who is more interested in whom, and who feels more threatened by whom.

In addition to gaze direction, we are also unconsciously busy checking one another's pupil signals. The black spot at the centre of the eye grows larger, or smaller, as it controls the amount of light falling on the retina, but there is also an emotional response that can interfere with this reaction. If we see something we like our pupils dilate more than they should and if we see something distasteful, the pupils contract more than they should. Intuitively, a man knows if a woman is falling in love with him by the exaggerated dilation of her pupils. He won't know *how* he knows, but he will know. If, on the other hand, she is pretending to be very fond of him, but her pupils are mere pinpricks, even in a dim light, then she is faking her interest. The pupils cannot lie because we have no conscious control over them, so, like the changes in gaze direction, they provide the eyes with important mood signals.

Another oddity of the human eyes is that their tear glands seem to be unduly active. Other primates do not weep copious tears, but humans do. This is more marked in the female, but strong men have been known to weep openly wherever their cultural rules allow it. In some societies it is considered a serious weakness for a man to weep in public. This is usually the case in those countries that have a strong military history, where the 'stiff-upper-lip' approach is considered a matter of male honour. But wherever such rules do not apply, adult males can be seen crying freely in public when some disaster has struck. Males who suppress their weeping are not, in fact, doing themselves any favours, because misery tears are known to contain stress chemicals that are discarded by the shed tears. So the act of weeping is in reality helping to lower the level of inner stress, a valuable benefit for someone who is overwrought. This is why we feel better after a good cry. Tough males who struggle to hold back their tears are robbing themselves of this advantage.

Temporary eye-strain is a common affliction of civilised man. The fact is that our eyes evolved to work efficiently at much longer distances than are usually encountered in modern life. Prehistoric men did not sit bent over desks or slumped in armchairs, poring over figures, reading small print or watching flickering images on screens. As hunters, their eyes were more concerned with images in the far distance. More effort is needed for the eye muscles to focus the eyes on a near object than a far one, so the close-vision urbanite can easily give himself muscle fatigue by spending hour after hour peering at a spot only a few feet in front of him. When we watch television, work at our computers or read a book, it is not just proximity that causes the problem but lack of variation in the depth of vision. This forces the eye muscles to hold a particular degree of contraction for an unnaturally long period. Our eyes may ache, but that does not mean we have damaged them any more than a man who runs a mile has damaged his aching legs. All they need is a rest. The solution is simply to look away from the screen or page occasionally and focus on some more distant object for a few moments.

Eye make-up has been confined largely to the human female during the course of history, but there have been some interesting exceptions. In ancient Egypt, high-status males also decorated their eyes, using green and black as their favoured colours. They painted their lower eyelids green and their eyelashes and upper eyelids black or dark grey. The green was originally made from malachite, an oxide of copper, and the black was a complex mixture called *kohl*, a black powder comprising 'burnt almonds, oxidized copper, a couple of different-coloured copper ores, lead, ash, and ochre'. To give this black eye make-up a creamy-smooth consistency, they added 7–10 per cent of fat to the powder, mixing it on a special slate palette.

Egyptian eye make-up was more than just a beauty aid. It is thought that it may have offered some protection from the glare of the sun. It was also believed to protect the wearer against the Evil Eye and, more realistically, against insects and disease. The claim that it acted as a disinfectant, as well as an insect repellent, appears to be justified. When French chemists analysed the contents of the Louvre's collection of four-thousand-year-old Egyptian make-up containers, they discovered that the compounds they had been using in ancient Egypt were also employed centuries later in the Greek and Roman civilisations, to combat infectious eye diseases such as conjunctivitis and trachoma.

In modern times, male social rebels, especially in the field of popular music, have occasionally worn eye make-up. Among the first heterosexuals to risk this were Mick Jagger and Keith Richards of the Rolling Stones, who wore eye make-up for performances of 'Jumping Jack Flash' in the late sixties. In the 1970s they were followed by performers such as David Bowie and Alice Cooper in a movement usually referred to as Glam Rock. This was in turn followed by Punk Rock, in which the male performers also used eye make-up, but now in an outrageously excessive manner.

In the cinema, Alex, the central character of Stanley Kubrick's violent masterpiece *A Clockwork Orange* (1972), wore elaborate make-up on his right eye, including huge, artificial eyelashes painted directly on the skin around the eye. As the young man was portrayed as a savage thug, there was no danger of this fancy eye make-up appearing effeminate. Instead it had a strangely chilling quality, combining fastidious cosmetic attention to detail with mindless brutality.

More recently, in the early years of the twenty-first century, there has been a return to conspicuous eye make-up by actors such as Johnny Depp in films where he portrays a swashbuckling pirate. Some reviewers criticised Depp for looking like a drag queen in what should have been a defiantly masculine role, but he defended his decision to use eye make-up, commenting: 'The kohl came from how athletes wear black [just below their eyes] for reflection. I started thinking about the tribes of northern Africa, and the Berbers who have been using kohl under their eyes for thousands of years, which is medicinal, and protects the eyes from sand and sun.' With these remarks, Johnny Depp takes us right back to ancient Egypt.

Unusual male eyes have sometimes been viewed as highly desirable. One of the first ever 'celebrity insurance policies' was the one taken out by Hollywood's silent movie star Ben Turpin, whose great success relied on his dramatically crossed eyes. In the early 1920s he took out a policy with Lloyd's of London that would give him the then large sum of $25,000 should his eyes ever become uncrossed. This might seem like an unnecessary precaution, but he was concerned that, because his eyes had only become crossed as a result of an accident when he was a young adult, a further serious accident might have the reverse effect. During his days as a slapstick, knock-about comic, he would always check his face in a mirror after he had taken a blow to the head, just to make sure that his famous eyes were still fully crossed.

Ben Turpin was not the only actor to make a fortune out of defective eyes. Marty Feldman, famous as Baron Frankenstein's creepy servant Igor (pronounced 'eye-gore'), also relied on his distressingly bulging eyes to make him famous. They were the opposite of crossed, they were divergent, caused by a hyperactive thyroid condition. There was a swelling of tissues behind the eyeball, a decreased range of motion of the muscles around the eyeball, and a pulling back of the eyelids. As a result, his eyes appeared to be about to explode at any minute. In slang terms, this outward squint is known as being 'wall-eyed'.

Another famous face with strange eyes belongs to rock star David

Bowie. It is usually said that he has one eye a different colour from the other, but this is not strictly true. What happened was that, when he was a fourteen-year-old schoolboy, he had a fist fight with a friend over a girl. The friend's knuckle, holding a school compass, caught him in the middle of his left eye and damaged the sphincter muscles there. Despite two operations, he was saddled with a permanently dilated pupil in the left eye. This means that, in a bright light, when the pupil of his right eye contracts, and that of his left eye does not, his eyes do seem to be two different colours. In reality both his irises are blue, but from a distance the enlarged left pupil makes that eye look much darker.

The infamous gothrocker Marilyn Manson is another performer who appears to have an eye defect, displaying one normal eye and one with a pure white iris. The myth was invented that he had achieved this look by using a heated sewing needle to 'vaporize the cornea'. But the more mundane truth is that he created the bizarre effect by wearing a stylised or decorated contact lens in his left eye. Manson now has a contact lens style named after him – the 'Marilyn Manson Look' – and he has created a fashion for fun contact lenses decorated with skulls, flags, blood, sun rays, dollar signs, hearts, flames and other motifs.

Finally, there are several gestures involving the eye. The best known is the wink, a deliberate, one-eyed blink that signals a shared secret between the winker and the winked at. The collusion is based on the idea that the closed eye, aimed at the companion, is keeping their secret, while the open eye, aimed at the rest of the world, is excluding everyone else from the momentary intimacy. Performed between friends it signals a moment of shared, private understanding. Performed towards a stranger it requests a shared intimacy that has yet to happen. In other words, between strangers, it becomes a flirtation signal. In books of etiquette, the wink is viewed as a vulgar gesture. In polite society women rarely wink; indeed, for some reason, many women find it hard to perform the action in a casual way, making the wink a predominantly male action.

Apart from the wink, the eyes can also send messages by widening to convey shock or surprise, narrowing into a threatening stare, blinking in disbelief, or glistening with emotion. They can be lowered in moments of modesty, flicked up to the heavens as a sign of exasperation, or held aloft in a display of mock innocence. A forefinger pointed at the eye, sometimes pulling down the lower eyelid, sends the message 'I am watching you' or 'You are being watched'.

Another widespread eye gesture is the eye rub, in which the forefinger rubs the eye or the skin near the eye. If the eyes are not sore, this action means that the gesturer wants to avoid his companion's gaze, but does not wish to admit the fact. In such cases the action is performed unconsciously and can be used as a telltale sign, either that the man is lying, or that he has become aware that the companion is lying. Either way, he feels uncomfortable and experiences a sudden urge to cut off visual contact. The rubbing action enables him to do this discreetly, providing an excuse for looking away.

A comic eye gesture involves placing the curled-up hands in front of one eye to create a tube, as if the gesturer is peering through a telescope. Popular in Brazil, this is usually done by one man to draw his male companions' attention to an attractive woman. There is also a two-handed version of this, where the curled fingers imitate a pair of binoculars, and the message is less voyeuristic, more a case of 'I can see you!'

Finally, there are two ways in which the eye is used when swearing an oath. In Saudi Arabia, the tip of the right forefinger is placed on the upper eyelid of the right eye. In Holland, the fingertips of the first two fingers are placed, one on each closed eye, and held there while the oath is being sworn. The message is 'May I be struck blind if I am not telling the truth'.

6. THE NOSE

The human nose is a proud nose. It protrudes from the face like no other. Monkeys and apes are typically flat-faced and this forces us to consider why we alone should have such a strange profile.

First and foremost, the bony projection of the nose helps to protect our eyes from damage. For the human male, increasingly involved in the dangerous pursuit of prey animals, this was especially important and it is no surprise to find that male noses are generally more strongly built than female ones. The cheekbone, the brow-ridge and the nose-bridge form a hard, bony triangle that surrounds the vulnerable, soft tissues of the eye. The force of any blow to that part of the human body will fall first upon this bony armour, which will take the brunt of the impact. One only has to look at the squashed nose of an old boxer to see how much damage it can absorb.

In one particular case, the presence of a bony nose indirectly helped to give the world some of its greatest works of art. When he was a boy, the Italian genius, Michelangelo, was struck violently in the face by a painter he had been teasing. The blow was so hard that his attacker later reported, 'I felt the bone and cartilage yield under my fist as if they had been made of crisp wafer.' After that, it was said of Michelangelo that 'his forehead almost overhangs the nose'. But the nose had done its job. Without the projection there to take the full force of the blow, the attack might have damaged Michelangelo's eyes so badly that his future as an artist would have been in jeopardy and we would never have been able to enjoy his masterpieces.

In addition to being bony armour, the nose is also a shield against unwanted substances entering the nostrils. The down-turned nostrils help to stop water entering the nose when diving and swimming, and in combination with copious nostril hairs and nasal mucus they assist in reducing the amount of dirt that enters the nasal cavities. One of the prices early humans had to pay for descending from the trees to walk on the ground was that they were much closer to wind-driven dust, and needed extra protection.

Another crucial change in our behaviour was that we started to talk and the resonance of our voices became more important. Anyone trying to talk with a bad head cold will know the unpleasant sensation of trying to enunciate their words clearly. When 'a cold in the head' becomes 'uh code en da hud' the importance of the large nasal sinuses to human speech is suddenly revealed.

Even more important is the non-stop air-conditioning function of the human nose. The lungs like to receive air that is warm, moist and clean, and it is up to the nose to ensure that the air entering the nostrils is, if necessary, heated up, dampened and cleared of any dirt particles before it reaches the windpipe. It is no accident that tribes living, for example, in the very dry desert regions of North Africa, have much taller, more prominent noses than those inhabiting the hot, humid regions of central West Africa.

So, to sum up, the human nose is a piece of bony armour, a water shield, a dust trap, a resonator and an air-conditioning unit. In addition, of course, it is the sense organ that allows us to detect the smells and stinks, the fragrances and stenches, and the perfumes and putrifactions of the world around us. It is true that most of our information about the outside world comes through our eyes and our ears, but there are moments when these let us down.

During quiet, intimate moments in the dark, when little is being said or seen, our noses become exceptionally sensitive to the body fragrances of our partners. These responses may be so primeval that we are not even consciously aware of the fragrances our noses are detecting, but they are powerfully arousing, nonetheless. The male will have a massive, unconscious response to the secretions of the female scent glands and these glands will magnify their output during sexual excitement. All this goes unnoticed by the higher centres of the male brain as he engages in sexual foreplay, but his lower centres will be on full alert.

At less pleasant moments, when our eyes can see nothing wrong and our ears can detect nothing threatening, our noses may send out a warning signal, saying all is not well here, be on your guard. This happens when we smell burning, for instance, but cannot yet see any fire, or when we gasp in horror at a vile stench before we can see its source. In this latter case we may hate the nose for exposing us to such a disgusting odour, and cover our face with a hand or a cloth, but in reality we should be grateful to it, because stinking chemicals and the odour of putrifying flesh are warnings of things that are potentially extremely harmful.

There is a feeling that all bad odours are somehow universally horrid, and smell the same to all forms of life, but this is not the case. If a vulture smells putrifying flesh it will find it immensely attractive and will hurry to the spot. For us, this is dangerous meat to eat and evolution has, as it were, tuned our noses to a different wavelength. Human excreta is also a dangerous source of infection and, again, our noses come to the rescue, giving it an unpleasant odour that encourages us to avoid contact with it. With some animals this is not the case. Rabbits, for example, have to eat their own droppings more than once to get the full nourishment from them. This process, called refection, is essential if they are to extract all the vitamins they need, so their night droppings – that are nasty and smelly to us – are appealing to them.

The sensitivity of the human nose is far greater than most people realise. We possess no fewer than five million scent cells, positioned high up in our nasal cavities, and capable of making extremely subtle judgements. True, we cannot compete with our dogs when it comes to scenting, since a pet dog has forty-four times as many odour-detecting cells as its owner, but we are nevertheless capable of responding to extremely dilute fragrances, given half a chance.

The reason we play down our nose power is because we have increasingly ignored and interfered with its operation. We live in towns and cities where natural fragrances are smothered, we wear clothes that cloy and sour our natural healthy body odours, and we spray our world full of scent killers and scent maskers. We even think of 'smelling' as somehow primitive and brutish – an ancient ability best forgotten and left behind. Only in certain specialised areas, such as those of the wine taster and the perfumer, is there any attempt to educate the modern nose and develop its full and extraordinary potential.

The nose is our main organ of taste as well as smell, and this requires explanation. The tongue is the true organ of taste, but it is very crude in its ability. It can distinguish only four qualities: sweet, sour, bitter and salt. All the other 'tastes' of our widely varied cuisine are in fact detected not on the eager surfaces of our slobbering tongues, as we munch and chew and gulp our meals, but on the small, odour-sensitive patches high up in our nasal cavities. Odour-bearing particles make their way there either directly through the nose as we bring the food to our mouths, or indirectly from the mouth itself. A meal may taste good on the tongue but it smells delicious in the nose.

Because of its association with bad odours, its link with primeval animal snouts and its tendency to dribble and run when we have a cold, the human nose has somehow become the joke organ of the face. We speak of smouldering eyes, delicate cheeks and sensuous lips with awe. But when we go out of our way to refer to the nose it is usually in some derogatory fashion. We have names such as schnozzle, conk, hooter, bugle and snoot with which to insult our nasal miracle of engineering and chemical detection. To be handsome, a human nose must not possess any particular character; it must be totally *without* any character. A brief glance at the faces of male celebrities in glossy magazines reveals that the very best noses are small ones. This state of affairs has become more marked in the present century and it is worth asking why this should be.

To understand the decline of the nose, in terms of aesthetics, we have to look back at the kind of proboscis with which we pushed our way into the world. As babies we all possess tiny, button noses. As we progress through childhood these small projections grow in proportion to the rest of the face and reach their maximum size in adulthood. So it follows that a small nose equals a young nose.

Add to this situation a cult of youth and the consequence is clear: the smaller your nose the younger you look.

For the female face the situation is acute, because men on average have larger noses than women. So, to be youthfully feminine it is doubly important to have a small nose. But even for men, today, a button nose or small, snub nose is favoured because it gives them the little-boy look. This makes them appear more juvenile and also less domineering, which suits the post-feminist mood. A man with a defiantly strong nose, jutting out before him like the prow of a ship, is not the stuff of which caring, sharing partners are made. The proud, aquiline nose of early movie heart-throbs, like John Barrymore, are strikingly different from the baby nose of today's Brad Pitt, for example.

Cosmetic surgeons report that the number of men requesting treatment is now on the increase, with 'nose re-shaping' at the top of the list. According to the American Society for Aesthetic Plastic Surgery, 24 per cent of nose jobs (rhinoplasty) today are carried out on men. What was once the sole preserve of women is now being invaded by large numbers of males who wish to resort to surgery 'to deal with image maintenance and change'. With nose jobs starting at $3,000, the images seen in the bathroom mirror must be causing acute concern, to say the least.

Reshaping the male nose is by no means a modern phenomenon. An Italian surgeon even published a book about it as long ago as 1597. He was promptly excommunicated by the Vatican for interfering with God's work. It was probably just as well because his technique left a great deal to be desired. His method was to take skin from the patient's arm and stick it on to whatever was left of his injured nose. Although this was, in principle, a sound idea, unfortunately the new tissue could easily be dislodged by a powerful sneeze.

In eighteenth-century India, where slicing off a man's nose was a common punishment for adultery, there was a thriving trade in making wax noses. Placed in position, these were then covered in skin taken from the victim's forehead. News of this technique reached Europe at the end of the eighteenth century and may well have led to the first attempts at modern cosmetic surgery in the West.

The most infamous case of modern male rhinoplasty is undoubtedly that of Michael Jackson. As a boy, his natural nose was broad and generous but he was determined to lead it on to the straight and narrow. He did this successfully, but then, as time went on, he seemed to be taking it further and further away from its original shape until it began to show signs of collapsing altogether. Jackson himself has admitted to having two nose jobs, but cosmetic surgeons who have studied his case have suggested that he may, in reality, have subjected himself to as many as thirty or forty surgical procedures over a period of twenty years. According to one unconfirmed report, a German surgeon recently took some cartilage from one of Jackson's ears and inserted it into his nose to prevent further subsidence.

Other famous male celebrities who have gone under the knife to improve their noses include Tom Jones and Ozzy Osbourne. Both have had their noses made narrower to good effect. Ozzy admitted that the operation had been excellent for his self-esteem, and that it had helped to boost his confidence. Singer Tom Jones had a pleasant, strong nose, but opted for a daintier, reduced one.

Despite recent increases in male cosmetic surgery, however, the problem will always be less acute for men than for women. A strong, prominent nose is still permissible and as we look around we can see a whole range of male nose types, from tall to hook to snub, each with its own special appeal. But because a strong nose emphasises and exaggerates a masculine feature, it remains a taboo for any self-aware female.

In earlier times a strong male nose was much more important than it is today. Indeed, it was almost essential for a man of social standing. Edgar Allan Poe went so far as to say, 'A gentleman with a pug nose is a contradiction in terms'. Napoleon Bonaparte declared: 'Give me a man with a good allowance of nose . . . When I want any good headwork done, I always choose a man, if suitable otherwise, with a long nose.'

This preference for a long nose almost changed the course of history in 1831 when Robert FitzRoy, the captain of HMS *Beagle*, took a strong dislike to the shape of Charles Darwin's nose. FitzRoy, a devotee of physiognomy, believed that it was possible to tell the character of a man by the shape of his nose and was not about to set sail on a long voyage with someone who might prove to be incompatible. Darwin lacked the sharp, aquiline nose that FitzRoy favoured, and the captain was on the verge of rejecting Darwin because his bulbous proboscis clearly indicated that he could not 'possess sufficient energy and determination for the voyage'. Luckily FitzRoy relented and Darwin was able to embark on the historic adventure that would lead to the theory of evolution. On his return, the great naturalist remarked wryly that FitzRoy had eventually accepted that 'my nose spoke falsely'.

Victorians were fascinated by the fashionable pseudo-science called physiognomy that linked facial details to particular kinds of personality. In the early nineteenth century there was a special branch of physiognomy dealing exclusively with the nose, called 'nose-ology'. Five nose-types were recognised:

1. The Roman or Aquiline Nose. This indicated a decisive, firm, energetic man.
2. The Greek or Straight Nose. This indicated a refined, art-loving character.
3. The Cogitative or Widening Nose. This indicated a serious, strong thinker.
4. The Hawk Nose. This indicated a shrewd, insightful, worldly man.
5. The Snub Nose. This indicated a weak, mean, disagreeable, insolent man.

This earlier dislike of the small nose was reflected in Edmond Rostand's famous play about Cyrano de Bergerac. The author has his hero declaim: 'My nose is huge! Vile snub-nosed ass, flat-head, let me inform you that I am proud of such an appendage, since a big nose is the proper sign of a friendly, good, courteous, witty, liberal, and brave man, such as I am.' Cyrano was not, as many people think, a fictitious character, but a real person, a French soldier and author of the seventeenth century who is reputed to have fought a thousand duels defending the honour of his extravagant proboscis. He was also the first author to describe space travel by rocket, a fact that would have appealed to Freud who would no doubt have read a great deal into the long nose/rocket equation.

The world's longest nose belonged to an eighteenth-century circus freak called Thomas Wedders. His nose measured an incredible 7.5 inches (19 cm) and was so extraordinary that he was able to make a living simply by exhibiting it to the paying public.

Today's longest male nose is no match for that, being only 3.5 inches (8.8 cm) long. It belongs to a fifty-seven-year-old Turkish builder from Artvin called Mehmet Özyürek, who won the second Annual National Long-nose Competition, beating off twenty-six rivals. The competition was started to 'make people feel comfortable with their appearance' and for Özyürek, at least, this has certainly worked. After his triumph he was quoted as saying 'There is not a person on earth with a more impressive proboscis. It is a masterpiece.'

Many famous men, from Charles de Gaulle to Jimmy 'Schnozzle' Durante, would surely have agreed with the idea that, nasally speaking, big is beautiful (and Schnozzle even took out a $50,000 insurance policy on his infamous schnozzola). Very large noses are not merely masculine, they are also phallic. The human male only has two long fleshy protuberances on the centre line of the front of his body. One is his nose and the other his penis. Symbolic equations between the two, either conscious and humorous, or unconscious and serious, are inevitable. They have occurred for centuries and were common in ancient Rome, where the length of a man's nose was said to indicate his virility. In this way the 'Roman Nose' became a special term of praise.

Among tribal societies the nose often takes on a totally different role, and one that is given considerable importance. The nasal passages are envisaged as the 'pathway of the soul'. Without realising it, today we still subscribe to this view when we say 'Bless you!' after someone sneezes. The blessing is bestowed because it was thought that the force of the sneeze might expel part of the soul, which would escape through the orifice of the nose. Later, in the Middle Ages, when violent sneezing often accompanied the first stages of epidemic disease, the blessing grew in meaning and it survives today as a relic saying.

In certain tropical tribal societies the treatment of a sick man included the blocking-up of his nose. This was to prevent his soul departing his body as a result of the illness. An Eskimo custom required mourners at a funeral to plug their nostrils with deerskin, hair or hay in order to prevent their own souls following the departed soul of the corpse. In the Celebes a man suffering from a serious illness would have fish-hooks attached to his nostrils, the idea being to hook his soul if it tried to escape, and thus prevent it from leaving his body. In many cultures the nose of the corpse itself is blocked to stop the soul passing through it and there are numerous other examples unearthed by anthropologists which reveal that there has been an amazingly widespread belief in the nose as the departure route for the soul. All are based on the idea that the soul is somehow connected to breathing – to the breath of life. Ordinary breathing through the nose, in and out, keeps a balance and nothing is lost. But with explosive sneezing and the tortured gasps of the dying, the traffic becomes one way, and superstitious precautions must be taken.

As already mentioned, another early belief about the nose was that its shape could be used to determine the true personality of its owner. There is only one element of truth in this physiognomic argument, and that is so obvious that it has little merit. If a particular man possesses an unusually ugly nose, of any shape, or an amazingly handsome nose, of any shape, then the appearance of his face in each case is liable to influence the behaviour of companions. Being ridiculed for an ugly nose or loved for a handsome one will, inevitably, have an impact on the developing personality of the nose's owner. An ugly, ridiculed boy will grow up into a different kind of personality from a handsome, popular one. To this extent, nose shape does become connected with adult character, but this is a very different matter from claiming that every subtle difference in nasal contour can be read with precision as an indicator of personality traits.

One important way in which the nose *can* be read is as an indicator of changing emotions. Like the whole of the face, the nose is equipped with muscles of expression and we can show our feelings by the movement and postures of our noses, at least to a limited extent. The nose is far less expressive than either the eye or the mouth, but it does have a number of specific signals to offer the onlooker. There is the wrinkled nose of disgust, the twisted nose of distrust, the twitched nose of anxiety, the constricted nose of distaste, the flared nose of anger and fear, the snorting nose of irritation or repulsion and the sniffing nose in response to a detected odour. This is a simplification, but it gives an adequate picture of the range of nose signals we have at our disposal. With the addition of snivelling, snoring and sneezing, this is about all the signalling the human nose can do. Complex moods may produce mixed expressions, but these are the basic elements.

We also make contact with our noses in a variety of ways. Using our hands, we may touch or rub our noses when we are being deceitful, pinch our nose-bridges when we are deep in thought over some conflicting idea or in a state of exhaustion, or pick our noses when bored and frustrated. These are all auto-contact signals that are signs of self-comfort. In one way or another they all signify that the nose-contacting male is momentarily in need of a little help and opts to provide it for himself with the reassuring touch of his own hand or fingers.

If we are asked a difficult question and do not wish to tell the truth, the hand often shoots up to the nose and touches it, rubs it, grasps it, or presses it. It is as if the hand makes an involuntary move to cover the mouth to hide the lie, and then moves on to the nose. The final shift from mouth to nose may be due to an unconscious sensation that mouth-covering is too obvious, something that every child does when telling untruths. Touching the nose, as if it is itching, may therefore be a disguised mouth cover.

Some individuals, however, report that they have felt a genuine sensation of nose tingling or itching at the very moment they have been forced to tell a lie, so that the action may be caused by some kind of small physiological change in the delicate nasal tissues, as a result of the fleeting stress of the deceit.

It should be emphasised that not all involuntary nose-touching indicates actual lying. It may, in a few instances, reveal that a person was considering lying, but then finally decided to tell the truth. What all cases of involuntary nose-touching do have in common is that, at the moment the action takes place, the performer is reacting emotionally to the situation being faced, even though outwardly they appear calm. The inner thoughts are seething, while a decision is made to lie or, with difficulty, to tell the truth. It is that inner turmoil, following a difficult question from a companion, which the nose touch reveals.

The pinching of the nose-bridge when deep in thought probably has a similar basis, with the nasal sinuses beneath the bridge causing mild temporary pain as part of the nose's response to stress. The action of pressing the bridge between the fingers would tend to relieve this pain or at least respond to its presence.

The male nose is a popular organ for making symbolic gestures. Females hardly ever perform these actions, in many cultures it being deemed improper for a woman to perform gestural actions of any kind. More than forty of these male symbolic gestures are known, many of them used only locally. Some of the more unusual ones are as follows.

In the nose circle, one hand is brought up to the nose, where it encircles the nose-tip. This hand-ring is placed on to the nose and is then rotated, clockwise and anti-clockwise, as if the nose is trying to insert itself deeper into the tunnel of the hand. This is meant to represent the act of anal penetration, with the nose as penis and the hand as anus. It is a North American gesture signifying that a man is homosexual and is usually employed as an insult. The same action can sometimes have a slightly different meaning. The hand still represents the anus, but the nose now stands for itself instead of for the penis. This gesture, known as brown-nosing, implies that someone is a servile flatterer, so fawningly anxious to impress his superior that he engages in 'arse-kissing', this action being caricatured as the pressing of the nose into the anus of the dominant individual.

Everyone picks their nose in private, but in Libya and Syria a deliberate nose-picking action is employed as an insult. A forefinger is thrust into one nostril and a thumb into the other. They are then flicked forward together, towards the insulted victim, as if flinging snot at them.

Nose-to-nose contact is the friendly welcome of the New Zealand Maoris and certain other tribal groups. This is usually referred to as nose-rubbing, but on formal occasions the nose-tips are merely touched. In origin, it harks back to the time when the nose was used to sniff the body of a returning companion. Although we are not always aware of it today, we are capable of identifying our loved ones and our close companions by their individual body fragrance. Greeting someone by sniffing them was done, not only to recheck their identity, but also to explore any changes in fragrance that had occurred during the period of separation. It has recently been discovered that our sensitivity to personal fragrance is centred in a small cavity inside the nose that acts as a specialised scent detector. We are not conscious of the odours it detects, but we nevertheless register them and remember them. The Maoris are not alone in employing this form of nasal greeting. Among the Bedouin, the nose-tips of two men are brought together and touched three times in quick succession as a form of friendly greeting.

For most people, wiping the nose is a simple cleaning or comfort action, but in East Africa it is also used as a specific signal. There it sends the message 'It doesn't matter' or 'It's not important'. It is performed in a stylised way, with the hand making a screwing movement around the nose, followed by a wiping action and a noisy exhalation. The gesture implies that a problem, like mucus from the nose, is best discarded and forgotten.

In Portugal and Spain a man may signal that he has no money by stroking his forefinger and middle finger down the length of the nose, from the bridge to the tip. Confusingly, in Holland stroking the nose indicates that someone is mean, but there only the forefinger is used. There may be an historic link between these two gestures, based on the fact that there was a Spanish presence in the Netherlands in the fifteenth and sixteenth centuries.

Tapping the side of the nose also has more than one meaning. But in most places it indicates that a man is 'sniffing something out'. Its message is one of alertness, but the exact form that this takes varies from place to place. One version signals that 'you and I share a secret that we must guard because others will try to sniff it out'. In the Flemish-speaking region of Belgium the message is

'I know what is going on, I can sniff it out'. Or it may be used as a threat, meaning 'I have sniffed out what you are up to and if you do not stop I will attack you'. In southern Italy, the nose-tapper is signalling that someone else is good at sniffing out the truth. The message is not 'I am clever' but 'He is clever'. An alternative meaning, also found in Italy, is that someone is nosing about and we must be alert to their presence. In parts of Britain, especially in Wales, nose-tapping is a direct accusation towards a person believed to be 'sticking his nose' into your business. Its message is 'keep your nose out of my affairs'. All these meanings of the nose tap are closely related, but their existence reveals the way in which a simple gesture can gradually start to alter its significance in different regions.

Thumbing the nose is a playful insult in which a man, or a boy, places the tip of his thumb on the end of his nose, with the hand held vertically and the fingers spread in a fan. The fingers may be held still or waggled back and forth. Both hands may be used, one in front of the other. This is an ancient gesture, at least five hundred years old, known throughout all of Europe and the Americas and in many other regions. It has one basic message that is understood everywhere: mockery. Its origin is obscure. It has been interpreted as a deformed salute, a grotesque nose, a phallic nose, a threat of snot-flicking and the display of an aggressive cock's comb, but its roots go so far back that nobody can be certain. Because it has such a long history it has acquired more names than any other gesture: to thumb the nose, to make a nose, to cock a snook, to pull a snook, to cut a snook, to make a long nose, taking a sight, taking a double sight, the Shanghai gesture, Queen Anne's fan, the Japanese fan, the Spanish fan, to pull bacon, coffee-milling, to take a grinder, the five-finger salute; in France: *Pied de nez, Un pan de nez, Le nez long*; in Italy: *Marameo, Maramau, Palmo di naso, Tanto di naso, Naso lungo*; in Germany: *Die lange Nase, Atsch! Atsch!*

In Libya, Saudi Arabia and Syria, touching the tip of the nose while saying 'On my nose!' indicates a solemn promise. This is a male Arab gesture that signifies a solemn commitment to do something. In origin, it is related to the ancient custom of touching the genitals when swearing an oath. In this instance the nose is acting as a symbolic substitute for the penis.

In Mexico, when the forefinger and middle finger of one hand make a vertical V sign thrust up against the underside of the nose, there is trouble brewing. This is an obscene insult, with the nose representing the penis and the finger V the vagina.

In southern Italy a man may place a finger on either side of his nose and wobble it from side to side. This means 'I do not trust you' and implies that something stinks and the gesturer is trying to shake the stench out of his nostrils. In the rest of the world a man would be more likely to wrinkle up his nose in disgust.

Male nose ornamentation is rare but not unknown. Modern males who enjoy mutilating their bodies to display courage and non-conformity sometimes have a hole pierced inside the nose, at the point where the septum cartilage ends. There, there is a narrow membrane of skin that quickly widens out into the soft, fleshy septum that separates the two external nostrils. Boring a hole in this membrane causes the minimum of damage and permits quite a large metal nose-ring to be inserted, rather like the ring put in a bull's nose to control his movements.

Some men prefer to insert a simple, plain, thick metal ring in their nose, while others favour circular barbells that end in small spheres. Some of these nose ornaments are fitted in such a way that they can be removed or modified for formal occasions, or when the dress code in a workplace demands unpierced bodies.

The nostril-stud, inserted into a fleshy nostril-wing, popular among women, is sometimes seen on men, but it is not common. Equally rare is the bridge-jewel inserted into a piercing at the top of the nose, where there is enough loose skin between the eyes to permit the boring of a small hole near the surface.

Because of the nose's role as a symbolic penis, these nose-piercings have sometimes been referred to as a case of displaced circumcision, but it is doubtful whether this is what the piercees had in mind at the moment they were being drilled.

Perhaps the most extraordinary form of nose decoration was that belonging to the sixteenth-century Danish astronomer Tycho Brahe, who lost the tip of his nose in a student sword fight and had it replaced with one made of an alloy of silver and gold. In the days of duelling, facial scars were worn with pride and his remarkable nose decoration would have seemed less odd than it would today, but it was an exceptional sight nonetheless.

7. THE MOUTH

Human lips are unique. A brief glance at other primates reveals that the lips of monkeys and apes are very thin. Those of humans are, by comparison, thick and fleshy. This is because we have retained the everted lips seen in tiny ape embryos. Apes lose these fleshy lips long before they are born, but we do not. We keep them all through life. This trend is more conspicuous in the human female, but even the adult male has lips that are conspicuous and a redder colour than the surrounding flesh.

These inside-out lips are useful to the human male when he is a baby sucking at his mother's unusually rounded breast, and again when engaged in oral contacts such as kissing, during sexual encounters later in life. Because the female lips are wider, fleshier and redder than the male's, it has inevitably followed that women have exaggerated this difference to create super-feminine lips. Men have nearly always been reluctant to decorate or enlarge their lips because this makes them look too effeminate, with the result that there is little to record on this particular part of the male body.

Attempts to introduce male lip cosmetics have not met with any great success, but there have been a few brave attempts to exploit this aspect of male grooming. One company introduced a male lipstick containing '. . . essential oils and aroma isolates . . . primarily as a mood enhancer and anti-depressant', but found to their surprise that it also acted as a potent sexual stimulant. Impregnated with the aroma of rose and jasmine it is now being promoted as an 'essential passion accessory', presumably because the female partners of the males who are brave enough to wear it are pleasantly shocked to discover that their males are, for once, smelling of roses instead of lager. This male lipstick is made more acceptable because it is transparent and does not redden the lips.

Other, more extreme forms of male lip decoration do exist, but they are not common, presumably because of the great sensitivity of this part of the body. Despite the intense pain involved, or perhaps because of it, some men have had words or pictures tattooed on the inside of their lower lips. These markings are not normally visible but loom into view if the lower lip is pulled down and out. The words chosen are, of necessity short. One man has PAIN on his lower lip, another has HARLEY and yet another has ROCK-N-ROLL. One of these men also had his lower lip pierced for the insertion of two metal studs, and it is hardly surprising that he is the one who chose the word PAIN to add to his strange display. One young man preferred an image of a skull and crossbones, while a more imaginative one chose to have two female breasts depicted. Precisely what appeal these forms of decoration have is not obvious. They do clearly demonstrate the ability to withstand acute pain and are sufficiently bizarre to appeal to eccentric companions. Perhaps that is enough.

Large lip plugs or plates are usually worn only by tribal women, but in some South American Indian tribes, especially in the Amazon region, it is the men who wear them. They do this to display their status, and the plugs are increased in size each year. The tribal elders demonstrate their high status by the impressive size of their plugs. In one tribe, the now extinct Abipon, who used to inhabit the region that is now Argentina, the men wore specially decorated plugs on the lower lip. These were made of wood like other lip plugs, but were then covered with silver or brass.

We know that labrets, as lip-piercing decorations are called, have been worn for at least 3,500 years by at least some Native Americans. The reason we can be sure of this is because, even though the metal studs and rings themselves become separated from the skull when the flesh of a corpse decays, and may be displaced or lost altogether, they do always leave a telltale sign on the lower teeth. By examining the skulls in ancient burial sites it is possible to detect the wear that mouth metal leaves by rubbing against the lower incisors. It is also possible to tell the sex of the skeletons and from these studies we do know that for a period of at least 1,500 years only males wore labrets, and only a limited number of those, suggesting that they were the mark of high-status men.

Later, among the Aztecs and the Mayans of Central America, it was again only the high status males who wore labrets. These were of the highest quality, often being fashioned from pure gold, with inset stones.

Passing through the lips and into the male mouth, we encounter a set of 32 teeth, comprising, from the midline, 8 incisors, 4 canines, 8 pre-molars and 12 molars. Because the male, on average, has slightly heavier jaws than the female, it follows that male teeth are slightly larger too, and a strong, healthy set of even, gleaming white teeth is today considered to be an essential feature of male sex appeal.

Modern dentistry leaves no excuses for the arrogant young man who claims that he needs no extra help to make him attractive to women. Yet despite this, dental statistics reveal that men are still reluctant to seek preventive dental care. As a recent dental report put it: 'One of the most common factors associated with infrequent dental checkups is just being male.' Women are much more likely to visit a dentist on a regular 'check-up' basis, whereas men tend to pay attention to their teeth only when a serious problem has already arisen.

Because of this male stubbornness, the average man will have lost five or six teeth by the time he retires. If he is a smoker, that figure rises to twelve teeth lost. Also, men are more prone to oral cancer. This reflects a modern masculine attitude that says 'it is effeminate to fuss too much over your personal hygiene. Women like their men to be rough and ready, not primped and prissy.' In reality, this is not true. Women may tolerate male mouths that smell of cigar smoke or bad breath, but secretly they hate this aspect of male bravado. In Alexandria, Minnesota, it is illegal for a man to have sex with his wife if his breath smells of garlic, onions or sardines, and a wife can legally demand that her husband brushes his teeth before having sex with her.

One form of oral abuse that has been popular with men for centuries, but is now on the decline, is the habit of smoking cigars or puffing on a pipe full of tobacco. The tobacco used in cigars and pipes leaves an unusually strong odour on the breath and the vast majority of women have, for themselves, always avoided this type of smoking.

In the 1970s, 34 per cent of men smoked cigars and 14 per cent smoked a pipe. Today only 4 per cent smoke cigars and 1 per cent smoke pipes. This reduction has been due to the link that has been publicised between smoking and cancer, but it is interesting to ask why any adult male should wish to fill his mouth with a smelly cigar butt or a dribbling pipe-stem. The answer lies in the infantile act of thumb-sucking. When infants need a little extra comfort, but their mother's nipple is not available, they will often replace it with their own thumb. Sucking this nipple substitute soothes them and sometimes they are offered a specially designed comforter or dummy for the same reason. Cigar-chomping and pipe-stem-sucking are merely adult forms of thumb-sucking, with the added advantage that something warm can be sucked out of these oral comforters, making them even better substitutes for the long-lost mother's nipple. And it is the soothing effect of these activities that gives them the edge over cigarette-smoking, the cigarette being too slim to make a good nipple substitute.

Many great thinkers and leaders, including Einstein and Churchill, have in the past been addicted to pipes or cigars, and the claim is that, for them, smoking was an aid to contemplation. Because they did not inhale the smoke, they ran little risk of lung cancer. A cigarette smoker is four hundred times more likely to suffer from lung cancer than a pipe smoker, for example. But they did suffer from three things – the possibility of oral cancer, the dismay of those around them who found themselves enveloped in billows of smoke and the possession of strong-smelling breath.

Pipe smokers have put forward an amusing defence of their oral obsession. Smoking a pipe, they insist, encourages patience and forces a man to '. . . slow down and consider. If some world leaders had been pipe smokers, the world would have been a different place . . . if Hitler had been a pipe smoker, he would have stayed an

artist. He wouldn't have gone into politics.' If this calming, paci-
fying effect of pipe-smoking really works, then the present trend to
outlaw all forms of smoking may give us future leaders of a kind
we may yet regret.

Deliberate modification of male teeth is uncommon but not
unknown. In certain tribes, the central incisors were removed to
make the canines look more fearsome. In other tribes the front teeth
were filed to sharp points, again to make the open mouth look
ferocious and threatening. More decoratively, in some cases the
teeth were adorned with precious stones or metals. These were
usually fixed to the teeth by recessing them in some way.

It is not known how long such dental embellishments have been
employed by human tribespeople, but it was recently discovered
that nine thousand years ago in Pakistan primitive dentists were
already capable of drilling perfect holes in human teeth. Sadly whatever filling was then applied, whether remedial or aesthetic, is not
known.

In modern times, inserting diamonds or gold in male teeth has
been a popular form of 'flaunting wealth' in certain American sub-
cultures, from the colourful world of the early jazz performers to
that of the later gangsta rappers. The first great jazz pianist, Jelly
Roll Morton, who died in 1941, boasted a single diamond fixed
into one of his front teeth. Sadly, he had to sell it during the
Depression of the 1930s, but he had already started a trend that
has, today, performed a crescendo. The modern rap culture is awash
with tooth jewellery.

Among rap lyrics, usually so difficult to understand, there is
a piece entitled 'Grillz', by the artist called Nelly, that sums up
the new male obsession with decorated mouths. Nelly is a man
– his name is short for Cornell, and the term grillz means 'mouth
jewellery'. He sings: 'Rob the jewelry store and tell 'em make me
a grill/Add da whole top diamond and the bottom row's gold.'
He goes on to explain that he literally: 'I put my money where
my mouth is, and bought a grill/20 carats, 30 stacks, let 'em
know I'm so fo' real.' And ends by explaining: 'My motivation
is from 30 pointers, VVS/The furniture in my mouth piece simply
symbolize success.' In other words, he admits that in wearing

diamonds on his teeth he is motivated solely by the desire to display his wealth.

Presumably the special appeal of this type of ostentation is that it is more resistant to theft than, say, an expensive wristwatch, a pendant or a bracelet. The use of the term VVS is interesting because it reveals a technical knowledge of diamonds. VVS stands for 'very very small inclusions' – and refers to the clarity grade of high-quality diamonds.

Because these mouth jewels have become so extreme they are no longer fixed into the teeth by drilling, but are instead slipped on like a dental brace. Most of them are now so large that they completely obscure the teeth and present a dazzling display every time the mouth is opened. Before buying a new grill a mould is first made of the teeth. This is then sent in to the supplier with the preferred design. And there are literally hundreds of designs available, from 'New Donut Set Cubic Zirconia Styles' to 'Blue Stones Pavé Set Polished Border'. This seems to have become the masculine version of the intricate world of female fingernail designs, although at the present time it has not spread out as widely across the various social divides. Certainly, all this careful oral attention puts paid to that old definition of a man's mouth as 'nothing more than a funnel into which he pours beer'.

Behind the teeth lies that multi-tasking organ, the male tongue. This wonderfully muscular piece of moist flesh can taste food, savour wine, lick stamps, aid mastication, refine speech, clean the mouth, make rude gestures and bring female partners to orgasm more quickly than the penis. The surface of the tongue is covered in about 10,000 taste buds, arranged in a special way. On the tip of the tongue we taste things that are sweet or salt; on the sides of the tongue things that are sour (acidic); and at the back of the tongue things that are bitter. We can, incidentally, also taste sour and bitter on the roof of the mouth, and sweet and salt on the upper throat.

We all enjoy the taste of a good meal, but for some food professionals detecting subtle variations in the flavour of a dish is taken to the level of an art form. For them, the possibility of losing their sense of taste is such a terrifying prospect that they have sometimes taken out insurance policies to protect themselves from such a disaster. Food critic Egon Ronay insured his taste buds for £250,000 and television chef Antony Worrall Thompson insured his tongue for £500,000. And the taste buds of one expert wine-taster were once insured for the staggering sum of £10,000,000.

These cautious tongue artists should be grateful that the barbaric ancient punishment of having the tongue cut out is no longer a serious risk. Even in ancient Rome it was an uncommon penalty and was rarely imposed. The Emperor Constantine did once order that an informer's tongue should be torn out by the root, and Leo I insisted that the murderers of a patriarch should have their tongues excised before they were deported, but these are isolated incidences. In his horror play *Titus Andronicus*, Shakespeare keeps alive in the popular imagination the link between ancient Rome and the brutality of tongue-cutting when he imposes this torture on one of the play's victims and then has the poor creature taunted with the phrase: 'So, now go tell, and if thy tongue can speak, Who 'twas that cut thy tongue . . .'

In modern times, only Uday Hussein, the monstrous son of President Saddam Hussein of Iraq, is alleged to have favoured this particular kind of punishment. A total of five unfortunate men are reported to have been mutilated in this way in Iraq, by Saddam's special commando units, usually for the crime of having used their tongues to criticise their President.

Besides its tasting, feeding, speaking and licking duties the human tongue also transmits several visual messages. These are primarily based on two infantile mouth movements: the stiffly protruded tongue of nipple rejection when the baby has fed enough, and the sinuously exploring tongue when the baby is searching for the nipple. In other words, there is the rejecting tongue and the pleasure-seeking tongue, and these are reflected in the ways in which adults display this normally hidden organ. A man who is concentrating hard on some personal task and who does not wish to be disturbed, pushes out his tongue like a sign saying 'Busy, keep away'. A man who wishes to be openly rude also sticks out his tongue as an unmistakable rejection gesture. In complete contrast, a man who is feeling randy and wishes to signal his desire for a sexual encounter, in which he may literally explore his partner with his tongue, uses the sinuous, curling movements of the searching, exploring tongue of infancy.

In addition, there is a powerful piece of body symbolism that sees the tongue as an echo of the male penis. Obscene mouth gestures often make play with the tongue as a symbolic penis and the lips as a symbolic vagina. A common form of sexual invitation, for example, is the slow protrusion of the tongue from between open lips, repeated several times to simulate copulation. And there is a sexual invitation performed by South American males that involves the slow wagging of the tongue from side to side within the half-open lips.

One of the most curious actions we perform with our mouths is the yawn. When we are bored or tired we often cannot help stretching our jaws open to the maximum and at the same time taking a deep breath. Anyone observing us is liable to find our yawning action infectious and in no time at all a whole group of people can be set gaping and covering their mouths with their hands. What does it all mean? The truth is that nobody really knows, although there have been some educated guesses. Any suggestion that it has to do with the intake of air is ruled out because fish yawn in water. Another possibility is that it is a rest-synchronising action similar to certain pre-roosting actions of birds. Viewed in such a way the yawn becomes more of a visual display, a signal that the yawner is about to go to sleep. Its infectious impact on others would then make good sense. Sadly for this theory solitary animals also yawn, so there must be some other factor at work. Could it be a specialised stretching movement involving chest and face muscles? Other stretching actions of the limbs and trunk often accompany it, and the net result is a slight increase in heart-beat which may be part of an attempt by the body to get more blood to the brain. Somehow this does not feel like the whole answer, so for the present yawning must remain an intriguing mystery.

Less of a mystery is the reason why people place their hands over their mouths when they yawn. This is usually said to be simply a polite way of concealing the inside of the mouth, dating from the time, before modern dentistry, when many adults had rotting black stumps of teeth. This is a plausible explanation but it happens to be wrong. The true explanation goes back much further, to times when it was believed that a man's soul might escape with his breath if the mouth were opened wide enough. The covering of the yawning mouth was intended to prevent the premature departure of the soul. It also prevented evil spirits from taking the opportunity to enter the body through the gaping orifice. Some religious sects believed that yawning was a ploy of the Devil and, instead of covering their mouths with their palms, they snapped their fingers as loudly as they could in front of their yawning mouths to frighten away the evil being. Even today, in some parts of southern Europe, Christians make the sign of the cross when they yawn.

Mouth-covering gestures when we are not yawning have different origins. For example, during conversation a person may raise the hand partially to cover the mouth, sometimes even keeping it there while speaking. This is a 'cover-up' in both a literal and a symbolic sense and occurs when the person making the gesture is trying to hide something from his or her companions. It is a signal of secrecy, evasiveness or deceit. The hand comes up to the mouth as if to block the words that might issue from the lips. It is wrong, however, to think that such a person must be antagonistic. It may be that all he or she is doing is concealing from you a home truth that might be hurtful to you.

That universally popular oral activity, the kiss, plays a double role today, being used both as a friendly greeting and as a form of sexual stimulation between lovers. In the greeting form, the lips are applied to different levels of the other's body according to the relative status of the kisser and the kissed. If equal-status kissers meet they exchange equal kisses, on the lips or the cheeks. If a low-status kisser meets a high-status being he may kiss the great one's hand, knee or foot, or the hem of his garment. In extreme cases he is allowed to kiss only the dirt near the feet. Such extremes are rare today. In a cultural atmosphere in which the ideal is that all men are created equal, it is possible to observe a dominant male being kissed on the cheek by someone who is of much lower status. These days, to be given the correct form of body-lowered oral greeting, such as a kiss on the hand, a man has to boast an unusually high status, like that of a pope.

Turning to a completely different kind of oral activity, spitting has a strange history. In ancient times it was considered to be a way of making an offering to the gods. Spittle, because it emerged from the mouth, was thought to contain a small part of the spitter's soul. By offering this precious particle to his supernatural protectors a man could enlist their aid in his endeavours. The danger in this was that if his enemies could collect some of the fallen spittle, they could work hostile magic on it and bewitch the spitter. For this reason some great tribal leaders employed a full-time spittle burier, whose task was to follow the Great One everywhere with a portable spittoon and bury its contents each day in a secret place.

The magical power of spittle led to its widespread use in the making of oaths and pacts, and spitting on the palms of the hands when making a bargain has persisted in some countries to this day. Fighters who spit on their palms before an encounter are also resorting to this early form of magical protection, although the action has long since been rationalised as moistening the palms to get a better grip on an adversary.

In Mediterranean countries where a belief in the Evil Eye became commonplace spitting was a defence against it. If someone afflicted with the Evil Eye passed by, people would spit on the ground to ward off the dangerous influence. In this way spitting changed from a sacred act into a gross insult. Eventually, spitting at someone became a symbolic act of intense hostility, which it remains to this day.

When it comes to projecting things from the mouth, sounds travel furthest. The normal range at which the male voice can be understood is about 200 yards, but the maximum distance at which it has been detected is an amazing ten and a half miles across very still water on a silent night. In certain mountainous regions of the world whistling languages have developed for communication across valleys between men working in the fields. On one of the Canary Islands, La Gomera, there is a whistled language called *silbo*, which is essentially whistled Spanish, the pitch and tone variations of the whistles replacing the vibrations of the vocal cords. There are four vowels and four consonants that can be strung together to form more than four thousand words. On a good day these whistled messages can be understood at distances of up to five miles.

The male larynx is one-third larger than the female. Male vocal cords are 18 mm long (0.7 inch), female cords only 13 mm (0.5 inch). Compared with the other great apes, the human animal has a much greater gender difference in loudness and depth of voice. This is something we take for granted and yet it is a significant development in the evolutionary separation of the sexes in our species. At puberty the voices of boys break and rapidly deepen with the adult male voice averaging 130–145 cycles per second. The voices of girls retain the higher pitch of childhood throughout the adult lifespan, at 230–255 cycles per second, one octave higher. The difference in pitch between the male and female laugh is even greater.

So why did this increased contrast develop? There are two separate questions to be answered here: why do men gain a deeper voice and why do women not do so? The deeper masculine voice gives the adult male a more frightening roar, snarl and shout. This can be used to intimidate human rivals, to drive prey or to scare off predators. When human males first began eating meat, before they became full-time hunters, they probably began by scavenging, getting together to drive off carnivore killers from their freshly killed prey. This required great courage and active cooperation. And they would certainly have benefited from a deeper roar to frighten away their powerful rivals.

The higher feminine voice gives the adult female a more juvenile quality. Along with several other features, such as less hairy body skin, the high voice of the adult female transmits signals to the male that make him feel more protective. By sounding like a child, the female can arouse the caring, paternal feelings of her mate and, in this way, improve her chances of survival when rearing his children.

Fiercely independent modern females may see this interpretation of their high-pitched voices as insulting but the fact remains that the primeval female, with her heavy parental burden, needed all the help she could get to protect herself and her offspring. If she could do this by employing juvenile characteristics to arouse protective, paternal feelings in her mate, then evolution would be quick to give her this advantage.

There are many regional gestures involving the mouth. Sometimes the same message is conveyed by slightly different actions as one moves from country to country. For example, the signal for 'Silence!' is usually thought of as the pressing of a raised forefinger to the closed lips, but in Spain and Mexico the message is more likely to be given by holding the lips tightly together between a forefinger and thumb. In parts of South America the signal is made with a thumb-tip moved across from one mouth corner to the other. In the Bible, silence is requested by the placing of the whole hand over the mouth. In Saudi Arabia the local variant has the forefinger held up near the lips, while the gesturer blows on it.

The signal for food is much the same the world over, with bunched fingers miming the act of pushing food into the mouth. But the signal for drinking has at least two forms. In most countries it simply involves the mime of tipping an imaginary glass up in front of the open mouth; but in Spain a different version dominates, where there is a local custom of drinking from a soft leather bottle which is held up high so that a jet of liquid squirts down into the mouth. To suggest drinking, the Spaniard mimics this action by raising the hand in the air with only the thumb and little finger extended. The other fingers are tightly bent and with them in this position the thumb is jerked downwards towards the open lips. A strange descendant of this gesture, found in Hawaii, is a legacy of the early Spanish sailors who visited the Pacific. Hawaiians employ the same hand posture, with thumb and little finger extended, as a friendly greeting. They no longer direct it to the mouth but instead wave it towards their friends. Most of them have no idea of the origin of this gesture which they make every day without thinking.

93

Oral gestures indicating anger include the Mediterranean speciality of flicking the teeth with a nail. The thumbnail of one hand is placed behind the upper incisors and then violently flicked forward towards the intended recipient. For some reason this gesture is on the decline. It was common as an insult in Northern Europe, including the British Isles, in the seventeenth century, but has since vanished there. Its stronghold today is Greece, but it is also well known in Italy, Spain and southern France. Among some Arabs it is also popular, but others prefer to demonstrate anger by simultaneously biting their lower lips and shaking their heads from side to side, miming the action of a dog killing a rat.

Praise is often indicated by putting the fingertips on the lips and then kissing them towards the object of affection. Smacking the lips together was originally intended as praise only of food but is now often used as a sign of appreciation of a tasty female. Vibrating the lips noisily while exhaling, which once signalled that something was very hot, is nowadays also used as an appreciative comment on a woman, indicating that she is considered to be hot stuff.

In Greece, there is a mouth gesture that is not always understood by visitors. A man may be seen to tap the tip of his forefinger several times on his lower lip, with his mouth slightly open. This is not, as it might seem, a request for food or for someone to shut up, but a signal that he wants to talk to them.

In countries where it is rude to point with a finger, a mouth point may be used instead. This consists of protruding the lips briefly in a particular direction and at the same time turning the head in the same direction. This is found as the common form of 'pointing the way' in a number of tribal societies, in the Philippines, in parts of South and Central America, in certain African cultures, and in Native Americans.

In Arab cultures, there is a characteristic Mouth Salaam gesture, when performing a friendly greeting. The first two fingers of the right hand are brought up to the lips. After briefly touching the lips, the hand is then waved forward lightly while the head is bowed. This is the minimum expression of the full Salaam, in which an Arab will touch his two fingers to his chest, his mouth and his forehead, in that order. Symbolically this formal greeting means 'I offer you my heart, my soul and my head' but among friends the reduced version, touching only the lips, is usually preferred. It replaces the Western handshake.

8. THE BEARD

The most conspicuous gender signal of the human male is the beard. In prehistoric times, before the invention of delicate implements such as knives and razors, adult men had long, dense hairs spreading out from their chins. This was more than a mere gender signal, it was also a species flag, clearly identifying the scary new primate that was starting to roam the land. Today the world record length for a beard is 17.5 feet (5.3 m). Although this figure represents a rare extreme, any normal adult male human who ceased to trim his beard for a decade would make an awesome impact as a hairy-faced primate, easily outclassing the various other mustachioed and bearded species. The combination of bushy facial hair and flowing scalp hair in the completely unbarbered prehistoric male must have been something quite novel on the animal scene.

The growth of human facial hair begins slowly at puberty, under the influence of male hormones. The typical adult female never sprouts more than a very fine 'peach-fuzz', which is invisible at a distance and can only be detected by unusually close scrutiny. By contrast, the adult male grows long hairs around his mouth, all over his lower face, his jaws and chin and his upper throat. These hairs increase in length by about a sixtieth of an inch per day, which means that if he stopped shaving completely he would be the proud possessor of a foot-long beard in just over two years. Generous and impressive beards are therefore the result of five or six years' growth, and would hang down to cover much of his chest region. No other biological feature, apart from the genitals, would make him look more different from the adult female, and no other species of animal can boast a chin appendage approaching this length.

The individual hairs of the beard differ in their texture from those on top of the head. The scalp hairs are finer and straighter than the coarse hair of the beard, creating a stiffer, more pronounced contour to the lower face. In other words, the thicker, wirier hairs of the beard transform the face more radically than if they were the same texture as the scalp hair.

There is also frequently a difference in colour. Some men with light scalp hair have dark beards and vice versa, as though evolution has been favouring a contrast between the two regions. Also, in older men, the greying of the beard often develops in distinct patches, rather than over the whole area. The most common form of two-tone beard has white patches beneath the mouth, with darker hairs around the sides of the face and on the throat. Possessing a feature like this, with differentiated zones of facial hair, some light and some dark, is something human males share with the males of many other primates. In some species, the border between the light and dark patches is much more clear-cut than it is in the human case, but in others, such as the chimpanzee, the lar gibbon and the lion-tailed macaque, it is remarkably similar. In each of these species there is a white patch of hair beneath the mouth and darker hair around this. The effect of this arrangement, in both men and monkeys, is to focus attention on the mouth region.

The primary function of the male beard has been hotly debated for centuries. A strong case was made for it as a natural scarf keeping the delicate throat region safe and warm. It was claimed that because the males had to go out on the hunt, exposed to all weathers, while the females and young stayed snugly at home in the tribal settlement, nature had provided the brave men of the tribe with an exclusive form of protection, keeping the underlying skin warmer in winter and cooler in summer.

There are two major flaws in this theory. Firstly, if the bearded males had developed clothing of some kind – animal skins, for instance – to protect the rest of their functionally hairless bodies, they could easily have added some sort of throat protector, if this was so important to them. If the beard is envisaged as developing before the advent of clothing then, of course, there is no sense in leaving much of the body naked while protecting the throat. If rugged, hunting males, out on the chase on a cold, Ice Age morning, were in need of hairy insulation, nature would surely have given them back their whole pelt of fur. Secondly, those human races most perfectly adapted to the colder regions of the world – the fat-lined Eskimos, for example – happen to display the least bushy beards. If beards were throat-warmers, they should be the hairiest of all.

If there is any truth in the temperature-control theory of male beards, it is more likely to have been in connection with cooling rather than warming. As a sweat-catcher, the curly beard hair might operate efficiently in very hot climates, increasing the cooling effect of sweat evaporation.

An alternative theory sees the male beard as nothing more than a display of masculine adulthood, a simple gender signal with no other properties. As a signal of masculinity, a kind of male flag, the beard's main impact is strongly visual, but it also appears to operate as a scent-carrier. The facial region possesses a number of scent glands and their products are retained better on a hairy face. During adolescence, when these glands are being brought into play for the first time, an excess of hormones can cause the skin disturbance we refer to as acne. It is a cruel twist of fate that the sexiest of teenagers are those with the most severe acne rashes.

As a visual signal denoting a mature, adult male, the beard helps to exaggerate the aggressive human posture of the jutting jaw. When we are angry, we project the chin forwards; when we are submissive, we pull it in and back. Males have heavier jaws and more protuberant chins than females, giving them a more aggressive facial shape, even in moments of complete relaxation. With beards added to this projecting display, they become even more juttingly hostile in their appearance.

It is because of this gender difference that hero figures are always portrayed with massive jaws and impressively proud chins. Unfortunate males with receding chins are insultingly dubbed chinless wonders and looked upon as effeminate. Conversely, strong-jawed women are automatically regarded as tough, rugged personalities. These responses persist regardless of conflicting evidence about the true characters of known individuals. Many a man with a weak chin, such as Frederick the Great, for example, has proved to be highly assertive in real life. But we cannot help reacting, at an unconscious level, to the ancient biological gender signals.

In earlier epochs, the beard was looked upon as the male symbol of power, strength and virility. It made men look mature and impressive. Men swore on their beards; beards were sacred; God was heavily bearded, and a shaven deity was unthinkable. The pharaohs of ancient Egypt wore false beards for ceremonial occasions, to demonstrate their high status and masculine wisdom. Even the female pharaoh Queen Hatshepsut wore a false beard to display her great power.

Although today bearded ladies are thought of strictly as circus freaks, in ancient times certain of the mythological Mother Goddess figures were shown as bearded in order to give them greater significance. Even the Christian Church was able to boast a bearded female martyr. Her name was Saint Wilgefortis, a virgin who met her death by crucifixion. Her legend tells how, when her father offered her in marriage, she miraculously grew a beard because she had vowed herself to virginity. Her masculine appearance ruined her marriage prospects and her enraged father then had her crucified. She was later venerated by women who wished to be liberated from abusive husbands.

Because beards were so important as a masculine signal, the rulers of early civilisations such as Persia, Sumer, Assyria and Babylon devoted an immense amount of time to grooming and decorating them. They used tongs, curling irons, dyes and perfumes. Their beards were coloured, oiled, scented, pleated, curled, frizzled or starched and, for special occasions, were sprinkled with gold dust and shot through with gold thread.

To many people in earlier centuries the idea of removing the beard was abhorrent, appalling, unthinkable. To them, having a shaved face was grotesquely unnatural. To lose one's beard was a desperate tragedy. It was the punishment meted out to a defeated foe, to prisoners and to slaves. To be clean-shaven was a disgrace.

To the deeply religious, removing the beard was often seen as an affront to God. Ivan the Terrible said: 'To shave the beard is a sin that the blood of all the martyrs cannot cleanse. It is to deface the image of man as created by God.' And there is an old Russian proverb that states: Shaving destroys the image of God.

In England, James Bulwer, writing in the seventeenth century, was equally adamant that shaving was an insult to God: 'The beard is a singular gift of God, which who shaves away, he aims at nothing than to become less man. An act not only of indecency, but of injustice, and ingratitude against God and Nature, repugnant to Scripture, wherein we are forbidden not to corrupt the upper and lower honour of the Beard . . .'

Despite these pro-beard rallying calls, it is clear that there have been certain groups of men, even in ancient times, who have preferred to face the world in a clean-shaven condition. As far back as 30,000 years ago there is evidence of the existence of some kind of primitive razor. The earliest ones were made of sharpened flint, and must have been extremely painful to use. Later, about three thousand years ago, metal was available and iron razors were fashioned. At some point – we do not know exactly when – in a completely separate development, the Aztecs of Central America were able to shave by forging razors out of volcanic obsidian.

In ancient Egypt around 300 BC it was considered, by the elite at least, to be nastily animalistic to have hair showing on any part of the body, and the priestly class shaved every three days. Egyptian technology made impressive advances and left behind evidence of beautiful razors, some plated with gold, others encrusted with jewels. The Egyptians did, however, value the male beard highly and this created a conflict. As already mentioned, the solution for the upper ranks was to shave and then wear a false beard, or *postiche*, on formal occasions.

The earliest examples of voluntary shaving always appear to have been limited to a specialised minority. For some, it was connected with the desire to display an 'enslavement to a god'. Young men would offer their beards to their deities as a sign of loyal submission. Priests might shave off their beards as a symbol of humility. But shaving on a more permanent and widespread basis seems to have been introduced as a military style in ancient Greece and Rome.

Alexander the Great is reputed to have instructed his troops to remove their beards in order to improve their chances of survival in close combat. It was considered that long beards would make useful hand-holds for the enemy, as one sometimes sees today in professional wrestling bouts. Roman soldiers were told to shave their beards for reasons of identification. Their shaven chins were easily distinguished from those of the hairy barbarians they were fighting.

In the Roman capital shaving became a fashionable trend and professional barbers were imported from Sicily to meet the demand. They employed a Roman razor known as a *novacila*, and it is said that there were many minor accidents. Their clients continued to visit them, however, because, as in modern times, they were the source of so much urban gossip.

Shaving became so favoured that it even started to feature in a *rite de passage*. Fashionable young Roman males would engage in a special Ritual of the First Shave, when their friends would gather to offer them presents, and they would place their shorn hairs in gold or silver boxes and present them as a gift for the gods.

The Roman general Scipio became a depilating fanatic and was shaved three times a day. Julius Caesar was also facially fastidious, but was too scared to allow servants to apply a razor to his throat, in case they had been bribed to slash him to death. He therefore subjected himself to the tedious and painful process of having his beard plucked out, hair by hair, with a pair of tweezers. His soldiers, in the meantime, had to rub their beards off with pumice.

As soon as there were two well-established alternative fashions, shaven and unshaven, it became possible for any social group or culture to make a statement of allegiance or rebellion by the way their males trimmed their beards. When the Christian Church split into two factions in the eleventh century, the Western clergy all shaved their faces to distinguish themselves from the Eastern Church. This difference has lasted a thousand years. There has never been a bearded pope, or a clean-shaven leader of the Greek Orthodox or Russian Orthodox Churches.

It has been claimed that the Norman Conquest of 1066 occurred because of a clerical error. Saxon spies are reputed to have mistaken clean-shaven French soldiers for priests and brought back a false report that led to defeat.

Sometimes the pendulum between shaven and unshaven swung in a new direction simply because of a leader's shaving habits. One French king had an ugly scar on his chin and grew a beard to hide it. Out of respect all Frenchmen of the period wore beards. One Spanish king was unable to grow a beard, so all Spaniards of the period went clean-shaven to honour him.

As the centuries passed, the decision to shave or not to shave became more complex, giving rise to all kinds of rules, regulations and punishments. Some religious orders were shaved and others were bearded. Some monks had to shave twenty-four times a year, others seventeen, and still others six times a year. In one monastic order, failure to comply was punished by six lashes of the whip. Some monks shaved one another but others, due to numerous casualties, employed professional barbers.

Some monasteries jealously guarded their shaven condition, insisting that, for ordinary lay people, it would be sacrilegious to copy a clerical fashion. There is a chilling record that one layman, who had removed his beard, was accused of imitating his spiritual superiors and had his eyes gouged as a punishment for his wickedness, a heavy price to pay for a naked chin.

According to one historian, the shaving of a beard was partly responsible for starting the Hundred Years War. It seems that Louis VII of France, finding himself in trouble with the Pope, removed his beard as an act of penance. His wife, Eleanor of Aquitane, who had never been able to study his face before, was so shocked by what she now saw that she shunned her husband and embarked on a number of affairs. When he later divorced her, she took her huge wealth and married the Duke of Normandy who later became Henry II of England. It was this shift of Eleanor's possessions, which weakened the French crown and strengthened the English crown, that is said to have encouraged the start of the war. If this is true, it is a remarkable example of a major disaster resulting from a single act of shaving.

Occasionally the social divide between the shaven and the unshaven has been a matter of class, as in Elizabethan times in

England when a tax was levied on beard-wearers. The lowest rate for sporting a beard was the then considerable sum of three shillings and four pence per annum. This restricted beards to the upper classes and made them into a financial status display.

In complete contrast, in other cases severe social ostracism was sometimes the fate of the bearded, which only the most stubborn and courageous could resist. In Massachusetts in 1830, for example, a bearded man in one town had his windows broken, stones thrown at him by children and communion refused by the local church. Eventually he was attacked by four men who tried to shave him by force. When he fought back, he was arrested and jailed for a year for assault.

Usually, such strict measures were quite unnecessary, the majority of men being happy to shave without any encouragement. Only in cultures that were temporarily dominated by military hostilities, or by a patriarchal attitude towards the male role, as in late Victorian society, did the beard become widespread and flourish for a brief spell as the social norm.

In modern times the strict rules of yesterday have all but vanished. Today most men can display a shaved face, a moustached face, a stubble-strewn face or a fully bearded face without fear of reprisals. The only exceptions are in those regions where extreme religious factions still impose rigid controls on physical appearance. In such cases, the well-dressed fanatic and the pious terrorist are nearly always fully bearded, displaying their bodies in a manner that is presumed to appeal to their belligerent gods.

According to some Muslim teachers, if you are a male who is a strict follower of Islam, you are officially obliged to wear a beard. This rule is now widely ignored, a fact that some of these teachers deplore: 'No one has called it permissible to trim the beard less than fist-length, as is being done by some westernized Muslims and hermaphrodites . . . Thus, a Muslim who shaves or shortens his beard is like a hermaphrodite . . . Shaving and shortening the beard is the action of non-believers.'

Another pleads with clean-shaven Muslims: 'From one brother to another, I say: Grow a beard, since it promotes Brotherhood in the real world . . . We know you think you are handsome without a beard, but who cares? What matters is how Allah sees you. And when you do grow a beard . . . please grow it correctly, i.e. Fist Length. That is the prescribed length and no shorter.'

When in power in Afghanistan, the Taliban took the hatred of shaved faces to extremes, severely punishing male Afghans for not wearing beards of the proper length. Some men, it is said, were killed for this crime, while others were severely beaten or mutilated by having their noses 'cut'. In parts of Pakistan, where the Taliban influence is spreading, barbers who shave beards are now threatened with death.

Curiously, there is no mention anywhere in the Koran of a rule requiring Muslim males to grow beards, and it is not clear how this idea arose. There is no doubt, however, about the ruling with male Sikhs. For them, the wearing of a beard is one of the five sacred insignia of their religion and cannot be modified to the slightest degree to adapt to the demands of modern living. This has caused problems in special circumstances. One elderly male Sikh in a Canadian hospital was, despite his protests, shaved by a nurse as part of normal hospital hygiene. The old man was distraught and, after he had been released from hospital, it was reported that he had to hide in the back of his temple, too ashamed to be seen by his fellow worshippers. He was quoted as saying that his forcible shaving had brought disgrace on his whole family. The family responded by filing a Human Rights complaint against the hospital authorities, from whom they received a grovelling apology and a promise that 'Our diversity officer has spoken to the nurse and all staff on the unit to educate them about the Sikh religion'.

Orthodox Jews are just as strict about beards. Their belief is that the Talmud regards the beard as the adornment of a man's face, which demonstrates maturity and piety, and is not to be shaved off. The Torah, they say, commands the Israelites 'not to mar the corners of your beards'. For this reason, the traditional Jewish beard is full and includes a moustache.

In modern, mainstream society, however, it is clear that, given freedom of choice, the vast majority of men (the best estimate is around 90 per cent) prefer to shave their faces daily rather than retain their natural beards. In the Western world today, outside the realm of religious extremism, bearded faces are largely limited to social rebels of one kind or another. Men who put a high value on their individualism or their eccentric lifestyle, or who belong to a macho-assertive group who wish to give the impression of masculine aggression and dominance, or who belong to a special category, such as the bearded warrior, the bearded sailor, the bearded hippie or the bearded artist, are the ones who refuse to succumb to the lure of the razor.

Typically, aggressive military beards are neatly trimmed, indicating that their owners are both domineering and well organised, while the shaggy hair of the social rebels is more likely to be unruly and straggling, reflecting their owners' lack of regard for social conventions and controls.

Political leaders and heads of state today are almost all clean-shaven because they wish to portray themselves as belonging to the social mainstream. Only one famous leader now displays a full beard and that is the Cuban President, Fidel Castro. His is an interesting case, because he also insists on wearing a military costume that is not a formal dress uniform, but a field battle dress. His iconic display says, 'I may sit in the Presidential Palace, but my rebel spirit is still out there, in the front line, where in the heat of battle there is no time to shave.'

For beard enthusiasts there are today many beard clubs that organise meetings and competitions on an international scale. Men with dramatically hairy faces gather to compare notes and to establish who has the most imposing appendage. There are separate moustache clubs, but they have now all been brought together under the aegis of the World Beard and Moustache Association. There are twenty-six clubs altogether, in eleven different countries, and they meet annually to crown a new World Champion.

Many of the contestants display splendid, large, natural beards, but others have taken the eccentric route with elaborate and complex designs that border on the narcissistic. For them, the growing of a beard has become a specialised hobby and a personal statement of extreme individuality. The high-maintenance demands and the general inconvenience of these remarkable facial ornaments mean that they are extremely rare.

Among the many beard types seen at these club gatherings, the following are perhaps the best known:

Chinstrap. With long sideburns coming forward and ending under the chin.

Chin Beard. A tuft of hair only on the chin.

Circle Beard. A chin beard and moustache connected by hair to the sides of the mouth, forming a circle.

Garibaldi. A wide, full beard with rounded bottom and integrated moustache.

Goatee. A tuft of hair on the chin resembling that of a billy goat.

Keldorn. A goatee combined with vertical extensions from the corners of the lips to the jawline, but no moustache above the lip.

Musketeer. A small, pointed goatee with a narrow, prominent, English moustache.

Royale. A tuft of hair under the lower lip, perhaps worn with a moustache.

Van Dyck. A thick goatee and moustache with upturned ends.

Verdi. With rounded bottom and slightly shaven cheeks with prominent moustache.

Viking. Covers the cheeks, chin and upper lip with a thick growth, with no skin visible through the hair of the beard.

But what of the majority, the 90 per cent of modern males who feel the urge to shave every morning of their adult lives? Why do they do it?

Given its significance as a masculine signal, the removal of the male beard today, when there are no cultural restrictions, appears both bizarre and perverse. The old rules of religion, class and fashion have all been forgotten, and yet a naked face has become the social norm, rather than a local oddity.

Why should so many adult males in so many countries want to throw away their most dramatic gender signal? It is an act of self-mutilation that does not protect the body or exaggerate natural masculine features. Indeed, it makes the shaved man appear less male. And it is not as though it is an easy matter, for shaving involves a great deal of effort. Apart from the cost of toiletries, consider the amount of time involved in the act of shaving.

If a man spends ten minutes shaving every day of his adult life (say, for sixty years), this amounts to a total of 3,650 hours of seemingly useless activity. It was Lord Byron who described shaving as 'a daily plague, which in the aggregate may average on the whole with parturition'. But he overstated his case: human parturition (he means gestation) takes an average of 266 days; a lifetime of male beard-shaving takes 152 days. But that is still 152 days that could have been spent more enjoyably doing other things. So there has to be some special benefit, or benefits, in having a clean-shaven chin, even if they are not obvious at first glance. Here are some possibilities.

A shaved face makes its owner look more:

Youthful. Juvenile males have no beard. In *Much Ado About Nothing* Shakespeare says: 'He that hath a beard is more than a youth, and he that hath no beard is less than a man.' For him, making men look juvenile was an insult to their adult condition, but in the modern world we have arrived at a time where the cult of youth is a powerful force in society, and both men and women strive to look as young as possible. Therefore shaving makes an adult male seem younger than he really is.

Hygienic. Beards can harbour dirt and disease, especially after slovenly feeding. A visitor to seventeenth century Ireland noted that there it was a common practice for men to use the beard as a napkin during meals, wiping their greasy fingers on it. Even a careful eater finds it hard to avoid trapping some liquids or solids in the tangle of hair beneath his lips. A shaved face is potentially both cleaner and healthier. There is an old Greek proverb that says: *A beard signifies lice, not brains.*

Friendly. Beards are jutted out when the human male adopts a primitive threatening posture. The thrusting forward of the male chin is a key element in an emotionally charged, aggressive approach and the presence of a large bushy beard exaggerates this thrusting action dramatically. In contrast, a shaved face appears to be far less assertive and much more friendly.

Expressive. Beards help to conceal subtle changes in facial expression and tend to hide the owner's personality, making him seem emotionless and furtive. People sometimes speak of a man hiding behind his beard. A shaved face is more exposed and therefore easy to read.

Fragrant. Beards act as a scent trap. As with armpit and pubic hair, there are scent glands in the facial region and saliva also carries a personal scent. The facial hair may easily become coated in these substances that may, after a while, become stale and smelly. A shaved face, however, is a potentially more fragrant face.

Smooth. Beards are rough to the touch. This makes them sexually abrasive. Some women like the masculine feel of facial hair when engaged in sexual foreplay, but if it is prolonged, the female facial skin can become reddened and sore. Therefore, a shaved face is gentler on the female's skin. But this only applies to a closely shaved face. Short bristles or stubble can do even more damage than a longer beard.

Open. Beards act as masks, concealing the finer details of the man's face. Thick, spreading facial hair tends to reduce individuality in men and make them look much like one another. It also imposes a psychological barrier between the bearded male and his companions. A shaved face appears more trustworthy. For this reason, lawyers are advised: 'If you have a client with a beard, no matter who is on the jury or who the judge is, make him cut it off.'

Self-aware. Beards require far less attention that clean-shaven faces. Although there are some bearded men who lavish a great deal of time on their facial hair, with carefully trimmed and precisely defined beards, there are many others for whom the wearing of a big, bushy beard is simply a lazy option. Their facial appearance implies a careless, undisciplined attitude towards life, whereas a shaved face transmits an air of efficiency, effort and precision.

Feminine. Beard removal inevitably makes male faces look more like female faces. Although few men, as they shave in the morning, are thinking to themselves 'This will make me look more feminine', it is an inescapable fact that this is one of the effects it has. Because a shaved face is less masculine, it reduces the visual impact of male threat displays and therefore helps to defuse male rivalry. For many people this softer look is part of the attraction of the shaved face, but for others, in the past, it was looked upon as an abomination. James Bulwer raged against the naked chin: 'What greater evidence can be given of effeminacy than to be transformed into the appearance of a woman, and to be seen with a smooth skin like a woman, a shameful metamorphosis . . . How more ignominious it is in smoothness of face to resemble that impotent sex?' Shaving, he roared, was 'a ridiculous fashion to be looked upon with scoffs and noted with infamy . . .'. He continued in this vein for some twenty-four pages.

In modern times, psychoanalysts also took a dim view of shaven faces. According to Freud, the study of dream symbolism reveals that the removal of the beard has a strange, unconscious significance. Because the beard is a dark, hairy patch out of which a long red tongue can be protruded, it is interpreted as a male sexual motif. Therefore, to Freudians, a shaved face is an emasculated face and a dream about being shaved is a dream about the fear of castration. It is no surprise that Freud himself wore a beard.

Conforming. At times when it is the general custom to display a clean-shaven face, wearing a conspicuous beard is the deliberate act of an outsider. The bearded pirate, the rebel fighter, the hairy artist, the bearded hippie, the unkempt inventor, these are the non-conformists who refuse to accept the status quo and who demonstrate this fact by making their faces as different as possible from the established members of polite society.

Respectable. Beards are often worn by those who have dropped down to the bottom of society, by tramps and vagrants, the homeless and the demented. They either have no facilities for shaving or are beyond using them. Their beards reflect their very low social status.

Evolved. Beards are hairy and animals are hairy, so having a hairy face makes the wearer of the beard seem more animalistic. Therefore shaving makes us even more of a naked ape and so more human. This was the main reason that ancient Egyptians started shaving themselves. Beards had to be worn in prehistoric times because there were no razors available then. This means that wearing a beard today has a primeval, primitive quality about it. Therefore a shaved face is unconsciously associated with a more sophisticated condition.

Personal. False beards are often worn as a disguise and it follows that a bearded face is more anonymous than a shaved one. Therefore a shaved face is more identifiable.

Moderate. Beards have often been associated with extreme religious movements. Hasidic Jews, for example, obey the Levitical Code that forbids the shaving of the beard. Inevitably, this religious connection has sometimes endowed the beard with a hint of fanaticism and zealotry. Therefore a shaved face has a more secular quality and a more middle-of-the-road flavour.

Reading through these different qualities, it becomes clear why shaving is so popular today. The clean-shaven face is much more suited to the gentler, more amicable 'new man' than is the bushy-bearded face. Indeed, the full beard is now seen more and more like a symbol of macho swaggering and potential aggression, a domineering beard. The clean-shaven face, on the other hand, is becoming more and more of a cultural appeasement display, by males who are no longer demanding physical dominance as a masculine right, but wish to be friendly. In other words, the clean-shaven male is making a visual statement that is a request for cooperation rather than competition. He is saying to his companion: 'I am reducing my level of masculinity in the hope that you will reduce yours. In this way we can work and play together without over-stimulating our ancient pugnacity.'

According to the pro-beard lobby there are several advantages to being bearded, even today. Wearing a beard, they say, makes a man look more virile, mature and dominant. It can make a weak face look stronger, a plain face more interesting and gives the impression of independence, wisdom and mystery. In purely practical terms, wearing a beard saves money on shaving equipment as well as saving time and energy spent shaving each morning. Men who eschew shaving do not have to worry about morning stubble or five o'clock shadow and they can avoid cuts, scrapes and nicks. A beard can conceal facial scars, compensate for baldness, and has the added appeal that it provides something to stroke, twiddle and even suck. A bearded man is one who refuses to follow the herd and bow to the dominant, clean-shaven fashion.

Happily, for the majority of men in the developed world today, there is the personal freedom to make up their own minds. The bigotry and cruel punishments of earlier epochs are a distant memory. Even the grubbiness of designer stubble is tolerated and, in some circles, admired. Anything goes on the hairy male face, and the result is a more varied, more individualistic and more intriguing social world for us all to enjoy.

The male beard can also be used as a prop to non-verbal communication. There are just a few gestures relating to the male chin and the bearded face.

Stroking one's chin as a sign that 'I am thinking about it' is a relic gesture dating back to the time when the beard was a traditional symbol of wisdom and when moving a hand through it was meant to indicate deep thought.

The beard as a symbol of slow growth and the passage of time is also exploited in many regional signals. In Germany and Austria, for example, the thumb and forefinger of the same hand are sometimes stroked down either side of a real or imaginary beard in a gesture which means: 'The joke you are telling me is so old that it has grown a beard.'

In France and northern Italy, the back of the hand may be flicked forward against the underside of the chin, as if rubbing against the beard. This is a gesture of boredom, carrying the message: 'Look how my beard has grown while I have had to endure this.' Done aggressively, however, this chin-flick gesture exploits a third symbolic property of the beard and means: 'I point my masculinity at you!' It is performed as an insult towards someone who is thought to be lying or is making a nuisance of himself. It is a modern gestural version of the primeval male beard-threat display.

Finally, for some males who take the step of dispensing with their primeval beard hair there is the possibility of exposing another masculine signal in its place. Individuals with unusually protruding jaws often display a small cleft or dimple in the centre of the chin, a feature that is widely considered to be appealingly masculine – so much so that Michael Jackson has had one surgically implanted. Today, if you lack one and you are prepared to travel as far as the Philippines, you can have one created for as little as $450. The most famous natural cleft chins are those belonging to Hollywood stars Kirk Douglas, Cary Grant and John Travolta.

9. THE MOUSTACHE

If the act of shaving makes an adult male look more friendly and cooperative, at the same time it makes him look more feminine. For some men this has aroused private anxieties. Such men want to look young and expressive and clean-cut, but they would also like to display a little masculine hair to flaunt their gender in a modest way. The answer is that famous compromise, the moustache. By wearing only the part of the beard that covers the strip of skin between the nose and the upper lip, they can have the best of both worlds. A mustachioed figure is clearly not female, and yet it retains the expressiveness and the cleanliness of the shaved face. On the juvenility scale, it looks mature but not old.

The macho moustache has been popular with many military men, for whom it is overtly male and yet strictly disciplined. They have waxed it, dyed it, preened it and twiddled it, often making it into the focal point of their masculinity. Each period has seen its own special style of moustache-shape, from the upward-twisted, sharp-pointed mustachios of earlier days, to the tiny, pencil-slim strips of the early movie heroes, to the extravagant handlebars of wartime RAF pilots.

In the British Army, from the nineteenth century to the middle of the First World War, all ranks were forbidden to shave the upper lip. On 6 October 1916 this regulation was abolished, and soldiers could at last display a completely clean-shaven face. It has been suggested that this change was due to the fact that some of the younger soldiers at the front were unable to grow decent moustaches, but an alternative explanation is that the conditions in the

113

trenches were so appalling, and infestations of lice so bad, that hygiene demanded the cropping of as much hair as possible.

In earlier military conflicts, the moustache sometimes became a badge of rank. Lower orders were only permitted a modest growth, but the higher you rose through the ranks, the larger and more flamboyant your moustache was allowed to become.

The moustache has sometimes been referred to as a symbol of obsessive but inhibited sexuality and masculinity. It has also been bluntly stated that a man with a moustache but no beard is a man with sexual problems. The basis for these remarks is simple enough. The wearing of a moustache reflects a need to demonstrate masculine gender, but the restriction of facial hair to this small carefully trimmed zone on the upper lip indicates restraint and tight self-control. There is an element of truth in this interpretation, but applied as a general rule it becomes meaningless. For many men the decision to wear a moustache is not a personal one but simply a response to a dominant local fashion.

For example, there are sometimes rapid shifts in the social significance of a moustache. A classic example of this has taken place in recent years in New York and San Francisco, where macho males have lost their moustache signals to gays. When homosexual males started to wear moustaches in the 1970s, the old-style, tough heterosexuals who also wore them found themselves being followed down the streets by eager would-be gay partners. Horrified, they quickly shaved off their once macho badges and a new phase of moustache display was born in these cities.

For some men, their moustache becomes a matter of pride, even obsession. As with beards, there are many moustache clubs whose members meet and take part in international contests to decide who has the finest whiskers. These clubs can be found in eleven countries (Belgium, France, Germany, Italy, Netherlands, Norway, Sweden, Switzerland, Ukraine, UK and USA). The annual championships take place in a different city each year.

One of the most famous of the moustache groups is the Handlebar Club. It was founded in 1947 in the dressing room of comedian Jimmy Edwards at the famous Windmill Theatre in London's Soho. The object of the club was, and still is, 'to bring together moustache wearers . . . socially for sport and general conviviality'. It still meets today, in the Windsor Castle pub in London, and arranges international meetings with such organisations as the Svensk Mustaschklubben (the Swedish Moustache Club) from Malmö, De Eerste Nederlandse Snorrenclub (the First Dutch Moustache Club) from Lelystad, the Snorrenclub Antwerpen (the Antwerp Moustache Club) and various other groups from around the world.

One noticeable feature of these super-moustache wearers is that some of them cheat. This is not to suggest that any of the members would ever dare to adorn their face with a false moustache, but rather that the real hairs employed are not all strictly upper-lip. Some of the most spectacular appendages are to be found on the faces of those men who groom their real moustaches out sideways to join up with their adjacent cheek hairs. These can be grown very long and encouraged to grow horizontally, so the bushy moustache appears to be greatly extended to a point where it can be seen even from behind the wearer, protruding dramatically on each side of his head.

An impressive example of this type of moustache is the one belonging to a sixty-two-year-old Turk called Mohammed Rashid. His facial hair is an incredible 5 feet 3 inches (1.6 m) long and has remained uncut for ten years. He is exploiting his unusual appearance by setting off on a round-the-world tour where he is charging people £3 a time to have their photograph taken with him.

These side-whisker moustaches are, in reality, a combination of moustache and beard. Even more unorthodox are those hairy-faced men who trim their whole beard so that it protrudes horizontally, to the left and the right of the face, making it look like a moustache that has slipped below the lower lip.

Among the whisker fanatics there is, in fact, a whole range of styles and shapes and each of these is given a separate category at the international contests. Some of the most popular ones listed are:

The Wild West. Large and bushy.

The Fu Manchu. With long, pointed ends drooping vertically downwards.

The Imperial. Thick, bushy ends curved upwards.

The Walrus. Bushy, hanging down over the lips, often covering the mouth.

The Poirot. Neat, with vertical, finely pointed ends.

The Nosebeard. Large and luxuriant, with hair growing down the sides of the mouth.

The Pencil. Narrow, thin, closely clipped. Also known as the Mouthbrow.

The Mutton-Chops. Sideburns connected by a moustache but with a clean-shaven chin.

The Zappa. Thick, with a small, square goatee under the bottom lip.

The Horseshoe. Full, with vertical extensions from the corners of the lips down to the jawline, resembling an upside-down horseshoe.

Some famous figures are remembered for their moustaches, including that well-known trio with toothbrush moustaches – Adolf Hitler, Charlie Chaplin and Groucho Marx. And then there is that unique appendage, the Salvador Dali moustache. Dali's face was most familiarly seen with the long, upward-sweeping waxed points of his moustache probing the air, as if they were some kind of strange radio antennae. When asked to explain the way he groomed his moustache, Dali replied that he used the pointed tips to receive messages from aliens in outer space. On another occasion, when he was accused of loving money, he claimed that he used the pointed tips to perforate dollar bills.

Dali is probably the only person ever to have had a whole book published about his facial hair. It was called, simply *Dali's Mustache*, and was decorated with photographs specially taken by Philippe Halsman. Halsman shows the infamous moustache in a number of strange poses. In one picture it is shaped like an 'S' and there are two paintbrushes running across Dali's face to complete the dollar sign '$'. In others, the whiskers are tied in a knot, or are pressed into service as a paintbrush, or become the hands of a clock, or appear on the face of the Mona Lisa.

While photographing Dali, Halsman asked him a series of questions and the interview is published in the book. One of his questions was: 'What do you think of Communist Growth during the last hundred years?' To which Dali replied: 'From the point of view of hair on the face, there has been a steady decline.'

Dali's eccentric preoccupation with his famous moustache knew no bounds. On one occasion Igor Stravinsky, the Russian composer, met the artist in the corridor of a New York hotel. He was surprised to see that Dali was carrying a little silver bell in his hand of the kind priests use when going to the house of a dying person to give the last rites. When Stravinsky asked why he was carrying the bell, Dali replied, 'I carry it and I ring it, so people will see my moustaches.'

One of the world's most dramatic moustaches belongs to the Indian Ramnath Chaudhary, who is happy to unfurl his 6.5-foot-long appendage to entertain a crowd. He claims that it used to be 11 feet long, but that after his father died he had to cut his hair and it is only now starting to grow back to its former glory.

There does appear to be a connection between India and long moustaches. Kalyan Ramji Sain managed one of 11 feet and 1.5 inches (339 cm). In Ahmedabad, in western India, sixty-year-old Badamsinh Juwansinh Gurjar had his moustache measured at 12.5 feet in 2004, after refraining from cutting it for twenty-two years. Sixty-one-year old Badamsingh Gurjar Khatana, from Kemri in Rajasthan, claims to exceed all these figures, with a moustache that is 13.5 feet long. This moustache, which has not been trimmed for twenty-six years, is carefully oiled every morning, but apart from that is left to itself. Its owner explains the Asian fascination with long moustaches by saying that men in his village take pride in moustaches as a sign of masculinity. With two wives, fourteen children and his massive moustache, Khatana's virility is not in doubt.

In terms of body language, there are two special moustache

gestures, the moustache wipe and the moustache-tip twiddle, which have the same meaning. They are preening actions implying preparation for courtship and are used as signals of arousal when an Italian male sees an attractive female and wishes to convey his feelings about her to his friends. The twiddling gesture dates from the days when waxed mustachios were commonplace and it has outlived them. Today, clean-shaven men will twist the imaginary tips of non-existent mustachios, but the significance of the gesture is not lost on their companions, even though they may be too young ever to have seen a waxed moustache.

Finally, there are two curious laws, still on the books, that affect moustache-wearers in parts of the United States. In the town of Eureka, Nevada, it is illegal for a man with a moustache to kiss a woman. This bizarre local restriction is apparently based on the idea that a moustache can harbour disease. The law reads: 'A mustache is a known carrier of germs, and a man cannot wear one if he habitually kisses human beings.' And in Alabama it is 'illegal to wear a fake mustache that causes laughter in church'.

10. THE NECK

The neck of the human male is shorter, thicker and more powerful than that of the human female. This difference probably evolved because it was important to the primeval male hunter to have a stronger, less easily broken neck when engaged in the potentially violent business of subduing a large prey animal. Today its greater strength is of much more limited use, although it can still make the difference between winning and losing to a champion boxer or wrestler. Terms such as bull-necked and swan-necked are still used to emphasise the difference between the necks of male and female.

The neck is a versatile piece of anatomy, providing the vital link between the all-important head and the rest of the body. Everything we eat or drink has to pass through its narrow girth, on its way from the mouth to the stomach. Every message from the brain has to pass through the neck to reach the rest of the body. Every breath of air we take in has to pass through the neck to enter the lungs. Every drop of blood that reaches the brain has to be pumped up through the neck from the heart. And every movement of the head, every tilt, twist, nod and shake, has to be orchestrated by the powerful neck muscles. So the neck provides a neat, compact housing for the oesophagus, the spinal cord, the windpipe, major arteries and veins, and complex musculature. No wonder that phrases like hanged by the neck, or put your neck on the block have such a sinister ring to them.

Apart from size, the most visible difference between male and female necks is the existence in men of a prominent Adam's apple, a visible lump at the front of the male throat that is much larger

119

than its female equivalent. This lump is caused by the underlying cartilage of the voice box and, for men who make their living as female impersonators, this is one of the most difficult masculine features to hide.

The origin of the term Adam's apple is not hard to guess. Early folklore maintained that the lump in the male throat was placed there to act as a constant reminder of Adam's original sin, the act of eating the apple offered to him by Eve. A piece of this apple was said to have become immovably lodged in his throat when he took his first bite of the forbidden fruit. In actual fact the word apple is not used in the Bible story of the Garden of Eden. It is a later invention, but the name Adam's apple has defiantly survived.

The extra protrusion of this anatomical detail in the male is due to the fact that, as established earlier, male vocal cords are 18 mm long, while those of the female are only 13 mm. The male vocal cords are also thicker than the female cords. As a result the male larynx, or voice box, that houses them is roughly 30 per cent larger than that of the female. It is also placed slightly lower down in the throat, which has the effect of making it even more prominent. The difference in the size of the voice box does not appear until boys reach their teens, when the male voice deepens, or breaks, sinking from a childhood pitch of between 230 and 255 cycles per second, to somewhere between 130 and 145 cycles per second. This equips the human male with an impressive, deep-throated roar that can be heard from a great distance.

For contact sportsmen a strong neck is more than an asset, it can help prevent a serious injury. 'Are you a pencil-neck?' booms a trainer in the tone of voice favoured by such men, 'If so, what are you going to do about that stick-like abomination protruding from your collar?' His answer is to exercise the four main muscle groups of the neck until the pencil-neck has become an invincible bull-neck, as wide, if not wider than the head above it. The muscle groups involved are the ROTATORS that rotate the head from side to side; the FLEXORS that move the head downwards, chin to chest; the LATERAL FLEXORS that tilt the head sideways, ear to shoulder; and the EXTENSORS that tilt the head backwards, eyes to sky.

Despite exercises to build a strong enough neck, there is always the risk of suffering a broken neck in a fall, or if struck hard on the head. This can lead to instant death or to severe paralysis. One of the most tragic public cases in recent years was the accident suffered by *Superman* actor Christopher Reeve. A keen horseman, and in superb physical condition, he was competing in a jumping event in 1995 when his horse stopped suddenly at the third fence and flung him headlong to the ground, breaking two of his neck vertebrae and damaging his spinal cord so badly that he never walked again. Indeed, he barely moved again and was unable even to turn his neck.

Mike Utley, an American football player, suffered a similar fate following a massive collision during a game in 1991, but managed to give a defiant thumbs-up gesture. Months of therapy gave him back some use of his lower legs and he went on to establish a foundation to help other paralysed sports victims. Ice hockey star Steve Moore suffered a similar fate, his broken neck ending his career, after being hit by a rival player.

These three examples of injuries to super-fit men give some idea of just how important the neck is and also how vulnerable it can be. But the damage done when the neck is broken is unpredictable, ranging from instant death to total paralysis to partial paralysis to little more than a pain in the neck. It all depends on how the spinal cord is damaged when the neck bones snap. In several cases sportsmen have completely ignored a broken neck and have continued as though nothing has happened. Kurt Angle even won a Gold Medal in the Olympics with a broken neck. And Manchester City goalkeeper Bert Trautmann famously completed the 1956 FA Cup Final despite breaking his neck fifteen minutes before the end of the match.

More recently, in 2003, the eccentric rock star Ozzy Osbourne gave the term Heavy Metal a whole new meaning when some of it landed on his head in the form of his quad bike. Luckily his broken neck healed and there was no paralysis, even though he was far from being a super-fit sportsman with finely tuned neck muscles. This emphasises what a lottery it is when the neck snaps, and how difficult it is to predict the outcome.

It also explains why, in the days when the 'humane' hangman plied his trade, in the first half of the twentieth century, great care was taken to ensure a powerful enough 'drop' to cause instant death. The long-drop method of hanging caused death by breaking the neck and severing the spinal cord. In earlier days, when public hanging was popular, the hangman used the short-drop to ensure that death was caused by strangulation. This was a much slower process, with greater entertainment value for the crowd of onlookers, who enjoyed the frantic struggling of the suspended victim, callously nicknamed the hangman's hornpipe. This brutal form of public theatre still occurs today in the Islamic Republic of Iran, where the victims are now hoisted into the air on powerful telescopic cranes.

In another Islamic country, Saudi Arabia, the male neck is still the focus of attention at public executions, but there the method is decapitation by the sword. In Europe, in earlier days, the use of an axe to sever the neck was so inefficient – it often took a number of blows to sever the head completely – that a gentle doctor by the name of Guillotin introduced a more efficient method in the form of a heavy blade that fell on to the back of the neck from a great height. The idea was that death would occur in a split second, but sadly this was not the case. The guillotine was so efficient that there was relatively little impact on the brain case, with the result that the decapitated heads went on living for up to seven seconds after the blade had fallen. This was the time it took for the massive drop in cerebral blood pressure to cause loss of consciousness, and audiences at the pubic guillotinings in France often reported seeing the decapitated heads performing facial expressions with their mouths, moving their eyes and blinking their eyelids.

In the arcane world of occult practices, such as voodoo in Haiti, it is believed that the human soul resides in the nape of the neck. This means that the neck is precious and needs special protection. This explains the wearing of necklaces before they became merely decorative. When first used they were thought to have the special function of protecting this vital part of the human anatomy from hostile influences such as the Evil Eye.

In modern times, the decorative necklace has become mainly the preserve of the female. Males have, over the centuries, switched largely to a different type of neck apparel, the cravat or necktie. In Freudian terms, the tie has always been seen as a penis symbol, with the brightness or dullness of any particular tie being interpreted as an unconscious indication of the wearer's attitude towards sexual practices. It has been pointed out that the V shape at the end of the tie is like an arrow pointing at the real penis, as if to remind us of its true significance.

To some feminists, the flaunting of a necktie was tantamount to boasting about the size of the male penis. One female boss in a German office even ordered her male civil service clerks to take off their ties, a ban that proved highly unpopular.

Others see the necktie as a symbol of bureaucratic control by the Establishment. One commentator complained that 'the necktie constantly reminds the wearer that his employer . . . has him by the neck. The necktie functions like the ring in the nose of the bull, the bit in the mouth of the horse. A necktie is a rope around the neck, a choke collar inviting enslavement that is constantly accepted . . .' Another echoed this view, stating: 'To those of us who wear it to work the tie is a burden . . . It represents the very essence of discomfort, as it applies light pressure to the very tube we all require to breathe, reminding us of our life-sentence to capitalism by tie-hanging . . . The very essence of conformity.'

More than mere decoration, the ties men put around their necks seem to have two very different, and conflicting, symbolic values – either brash phallic displays or servile badges of conformity. It follows that to remove your tie and wear an open-necked shirt is either an act of modesty, or of rebellion, or perhaps just a desire to be comfortable and give one's neck a greater freedom.

Male necklaces do occasionally make a comeback, but their use is usually limited to special types of male, such as the 1970s Medallion Man, the 1990s Bling Bling Man, and the twenty-first century Metrosexual. The Metrosexual Male has been described as 'a heterosexual male who is in touch with his feminine side'. A modern version of the old-fashioned dandy, he is concerned with his appearance to the point of narcissism, and wearing jewellery is a part of this trend. This includes necklaces, but they are not usually as vulgarly dramatic as those of the earlier Medallion Man or Bling Bling Man.

The definition of the 1970s medallion-wearer was given as 'a man, usually an older man, who dresses in a way that he thinks women find attractive, often wearing an open shirt in order to show his chest and a lot of gold jewellery'. Bob Guccione, the founder of *Penthouse* magazine, was the epitome of Medallion Man, before the style sank into caricature and ridicule.

The Bling Bling tradition, growing out of the American rap and hip-hop culture, favoured flashy pendants with designs such as 'Sterling Silver Dogtags' and 'Iced-out Crosses'. The term Bling Bling was coined by a New Orleans rapper back in the late 1990s and gained national awareness with a song entitled 'Bling Bling'. The wearing of this type of showy, expensive jewellery has since spread to the world of professional boxing and the infamous promoter Don King 'owns assorted diamond-studded crucifixes . . . and two famous crown-logo necklaces, the larger of which has a four-carat-diamond centre-piece and 36-inch necklace with bullet links fashioned from diamonds and white gold'. If necklaces really do have protective powers, as the ancients believed, then Don King must be indestructible.

Another form of neck decoration originates not in the jeweller's shop but in the tattooist's parlour. These permanent markings are rare, however, and the reason is obvious enough. Tattoos on arms, chests, backs and legs can be hidden by clothing at times when their owner does not wish to display them. Neck tattoos are always going to be peeping out above a collar or beneath the hair, even on occasions where they might be inappropriate.

Perhaps the most famous neck tattoo is the winged cross on the back of the neck of footballer David Beckham. It was meant to be a talisman that would protect his children, but he must have been dismayed by the press reaction when he first displayed it in public. One newspaper labelled him a yob, while another said he looked like a Hell's Angel biker gang member. Before long his cropped hairstyle changed to a much longer style that soon covered the offending design. Only much later, when the fuss had died down, did he revert to a cropped style.

The neck has also become the focus of certain bizarre sexual practices. Adventurous males, exploring the darker side of erotic rituals, discovered that, by applying pressure to the large carotid arteries that run up the side of the neck, carrying blood to the brain, a subject could be made dizzy and confused, and easy prey to sexual suggestion. What was happening, of course was that the subject's brain was being deprived of oxygen. In this state the victim allowed sexual attentions of a kind that they would have resisted had they been fully in possession of their senses.

It is a dangerous practice, of course, because if it is pushed a little too far it causes death. Because of this, one might expect it to be confined to the savage world of rapists, but, unfortunately, it has some pleasantly erotic side effects if it is carefully limited in its application, and is therefore more widespread. In fact, today, partial asphyxiation is used by a growing number of young men to enhance their sexual pleasure. Putting a noose around their partner's necks, or their own neck, and then tightening it, causes 'hypoxic euphoria', a condition that leads to loss of self-control, giddiness, light-headedness and exhilaration. This euphoric condition is said to intensify sexual pleasure at the moment of orgasm, which is what gives it its irresistible appeal to those who like to live dangerously.

Some men, once the practice takes a hold on them, find that eventually they can *only* become sexually excited when being semi-strangled or semi-suffocated. This is a condition known officially as Asphyxiophilia (or, in slang, as scrafing) and it frequently leads to death. This is because, repeated often enough, the chances of it going too far, just once, are very high.

Hundreds of young men die every year in this way. A notable recent case was that of the rock star Michael Hutchence who was at first thought to have committed suicide by hanging himself with his leather belt from a hook on his Sydney hotel bedroom door. But as he had no suicidal tendencies, left no note, had made plans for later in the day, and was naked at the time of his death, the general conclusion was, as one report put it, 'that his death was the result of accidental autoerotic asphyxiation, a dangerous form of masturbation'. A few years ago a Conservative MP was found dead in similar circumstances. He had died of asphyxiation from an electrical flex tied in a noose around his neck, he had a black bin-liner over his head and his body was naked except for stockings and suspenders, features that were 'consistent with auto-erotic sex practices'.

Turning to the body language of the neck region, there are two kinds of signals that are transmitted from here. The first involves the hand, or hands, making contact with the neck, and the second involves a contraction of the neck muscles to create a head movement or special head posture. Some of the more interesting hand-to-neck actions are as follows:

In the neck clasp, one hand is brought up to contact the side of the neck, just behind the ear. It is an act of self-comfort when something has suddenly gone disastrously wrong, the gesturer clasping himself as though giving himself a consoling hug. It is widespread, but is most commonly observed in Jewish communities.

In the Middle East, if a man lightly taps the back of his neck with his hand, it signifies that he thinks someone is homosexual.

Gently flicking the neck is an Eastern European gesture, done by one man on the back of another man's neck. The forefinger is flicked lightly against the skin as an invitation to come and have a drink, but it is only used between old friends. It is considered rude to do it to a stranger.

The throat-cutting gesture is a threat signal, in which a man does to himself what he would like to do to someone else. His stiff forefinger is drawn horizontally across his own neck, mimicking the action of a knife cutting someone else's throat. There is also a version using the edge of a flat hand. This is popular in television studios, when time has run out, to indicate that a speaker must stop immediately or he will be cut off.

Grasping the throat is another action in which a man does to himself what he would like to do to someone else. The exact message varies from place to place. In many the action means 'I will strangle you'. In others, however, the same action may be used to indicate suicide, either by the gesturer himself or by someone else. In Italy, it is more likely to mean 'I am fed up to here' and 'I have had enough.' In South America, it is a signal that someone has been caught and will go to jail, or that what is being done right now could lead to imprisonment. In North America, it is used by sportsmen to say that they have done badly. The American Red

Cross have also suggested the use of this gesture as an official sign to indicate that a man is literally choking. If he has a piece of food stuck in his throat, he is supposed to make the self-choking gesture with his hand, but the problem is that, since this action has so many different meanings in different parts of the world, he may not be properly understood. While onlookers are trying to interpret his frantic gesture, he may solve his problem by turning blue, a signal that few people would misread.

In addition to these hand-to-neck gestures, there are some that do not involve the hand, but in which the muscles of the neck are tightened in different ways to create head movements such as nodding, shaking, bowing and twisting. There are more than twenty of this second kind and a few of the more important ones are as follows:

In the nod, the neck moves the head vertically up and down one or more times, with the up-and-down movements of equal strength, or with the down elements slightly stronger. This is the most common and widespread action of assent or agreement. Although in some cultures there are other actions signifying assent, it remains true that whenever nodding occurs it has the same meaning: it always means 'Yes'. Nodding has a global distribution and early travellers found it in remote tribal societies that had not previously encountered Western influences.

Two suggestions have been made about the origin of the head nod. The first sees it as a modified version of the bow and interprets it as a highly abbreviated submissive body-lowering device. By saying 'Yes' a man is momentarily submitting to the other person.

The second suggestion relates the head nod to the infantile act of feeding at the breast. The nodding movement is seen as part of the pattern of accepting the breast, when the baby's mouth is searching up and down, feeling for the nipple. By contrast, rejecting the breast is characterised by the baby jerking its head sideways or upwards and this is said to explain why the negative head signals of adults involve sideways or upward movements.

In cultures where some other signal is used for 'Yes', this is usually an alternative to rather than a replacement for the head nod. In parts of Sri Lanka, for example, the head nod is used when responding to a factual question, but another movement, tilting of the head from side to side, is made when agreeing to a proposal. Both say 'Yes', but they are a different kind of 'Yes'.

In the head bow, the head is lowered into a head-down posture and then raised again. Unlike the nod it is always a single down-and-up action and the movements are stiffer and more deliberate. This head bow appears to be an almost worldwide greeting signal. In origin it is clearly a minor version of the general body-lowering of a submissive individual. However, although its basic message is 'I lower myself before you', it is not restricted to subordinates. When performed by equals, or by dominant individuals, the message is expressed in its negative form as 'I am not going to assert myself', which becomes generalised as 'I am friendly'. In strength, the action varies from an almost imperceptible bob of the head, to a dramatically exaggerated snap of the neck. The main cultural difference appears to be in the stiffness of the action, the Oriental head bow being much softer than the Germanic version.

Tossing the head sharply back is the opposite of the head bow, the head going up instead of down. It is used in several quite different ways, sometimes leading to confusion. Its most widespread occurrence is as a distant friendly greeting, performed right at the beginning of the encounter before closer interaction has taken place. Its message is 'I am pleasantly surprised to see you'. Surprise is the key factor here, the head toss representing a highly modified form of the full startle response. That the head toss and its opposite the head bow can both be used as greeting gestures is explained by the fact that the toss occurs at a distance and the bow at close quarters. Also, the toss is a gesture of familiarity, the bow of formality.

A second use of the head toss is as a signal of understanding. In this role it occurs during a conversation at the moment when someone suddenly sees the point of something and exclaims, 'Ah yes, of course!' Again this is a moment of surprise. It is as though the person concerned has suddenly started up or jerked backwards in a fleeting moment of alarm. The surprise is quickly dissipated and the head returns to neutral, leaving only the rapid head toss as a tiny intention movement of retreat.

In Greece and neighbouring countries, tossing the head back is a way of saying 'No!' This Greek 'No' gesture puzzles many visitors to the Eastern Mediterranean when they first encounter it. If they ask a civil question and in reply receive the sharp upward flick of the head, they imagine that they have irritated the gesturer, but cannot understand why. The reason for this is that in other parts of Europe there is a common irritation reaction, widely understood, in which the head jerks upwards, the eyes gaze upwards and the tongue is clicked. This transmits the message 'How stupid!' and the Greek 'No' gesture looks so similar that to a visitor it seems like a criticism of his enquiry. But to the Greek it simply means 'No' and carries no impoliteness.

In origin the Greek negative Head Toss appears to derive directly from the primitive breast-rejection reaction of the baby. Attempts by parents to spoon-feed an infant who is not hungry can easily produce a similar upward flick of the head and it is easy to see how this could grow into an adult negative gesture.

The more widespread way of signalling a negative is, of course, to shake the head from side to side. This also originates in infantile food refusal, either at the breast or when being spoon-fed. It has been suggested that in certain countries the head shake signifies 'Yes', but this is based on poor observation. The lateral head movement that sometimes indicates an affirmative involves tilting the head from side to side, rather than rotating it.

Turning the head sideways, away from a point of interest, is basically a protective device, but as a specific, deliberate gesture it is used as a 'cut' or rejection. In this form it is a silent insult telling someone that you are disengaging from social contact with them. But unless the movement is done well there is the problem that the studied insult might be mistaken for a shy hiding of the face. It requires that the insult movement be done boldly and exaggeratedly. Unfortunately this gives it a somewhat pompous flavour that has curtailed its use in modern times. It was, however, extremely popular in the nineteenth century. Its main use then was to keep the *nouveaux riches* in their place. The industrial revolution had dramatically increased the wealth of the middle classes, whose dearest wish was to use their new position to elevate themselves in social rank. This the upper classes resisted by introducing 'the cut'.

Etiquette books explained the technique: 'The individual should be allowed to see that his approach has been noticed, whereupon you turn your head away.' Today, although the formal cut is extremely rare, it is still common at moments of intense petulance during family quarrels.

Tilting back the head, and holding it in the tilted position, is the nose-in-the-air posture of the snob or the unusually assertive individual. The emotions of the man with the head tilted back range from smugness and haughtiness to superiority and defiance. This is essentially the posture of challenged rather than calm dominance. The success of the movement depends on the fact that the level of the eyes is slightly raised, giving the illusion of increased height. Short men who are forced to look up at their companions when talking to them often give the impression of being pompous when they are not, because it is natural for them to adopt the tilt-back posture of the head for purely physical reasons.

If the eyes are not looking at the companion but are also raised up, or perhaps shut, then the message is quite different. Heads tilted back in this way belong to men who are in agony or ecstasy, experiencing a massive dose of pain or pleasure. They are suffering from sudden overstimulation and their response is to cut off from the world around them. If the sensation is extreme enough, then, regardless of the exact nature of the emotion, the head tilt-back occurs and, with it, a temporary release from the source of the input.

Cocking the head to one side derives from a childhood action in which a child rests its head against the body of its parent when seeking comfort. If an adult male uses this action as part of a flirtation, the tilted head has about it an air of pseudo-innocence. The message says, 'I am just a little boy in your hands and would like to rest my head on your shoulder like this'. If used as part of a submissive display, the gesture says, 'I am like a child in your presence, dependent on you now as I was when I laid my head on my mother's body'. In this submissive form, the head cock is popular among old-fashioned shop assistants and obsequious headwaiters who wish to increase the feeling of superiority in their clients.

These few examples of the wide variety of head postures and actions reveal just how complex the human neck muscles are and how much flexibility they offer to the head they support. Compared with an animal such as the crocodile, or the rhinoceros, human head movements are amazingly subtle and expressive and, consciously or unconsciously, we learn a great deal about the mood and intentions of our companions from them.

Finally, this narrow yet powerful part of the male anatomy has donated several terms to the English language. These include, for example, redneck, rubberneck and necking.

'Redneck' is an insulting name given to 'a lower class white person from the southeastern states of the USA'. It refers to the fact that he is usually employed in outdoor manual labour, with the result that the sun has given him a reddened neck.

'Rubbernecks' was originally the name given to tourists visiting New York and craning their necks to look up at the tall buildings. It spread from there to describe anyone twisting their neck to satisfy their curiosity – as happens when motorists are passing a motorway accident. Finally, 'necking' was the term given to the actions of courting couples back in the 1940s and 1950s, when there was still a widespread taboo on the more advanced forms of intimacy. Couples typically remained clothed and kissing the neck was about as far as oral exploring could go.

11. The Shoulders

The main function of our shoulders is to provide a strong foundation for our multi-purpose arms. Ever since our ancestors adopted an upright way of life our front legs have become increasingly versatile, and our shoulder girdle, or pectoral girdle, has had to serve that versatility by becoming more flexible. The shoulder blades are capable of movements through about 40 degrees and, with their powerful muscles, can help the arms to swing, twist, lift and rotate in an amazing number of ways.

Each shoulder blade, or scapula, is connected to the front of the body via the collarbone, or clavicle, and to the upper arm, or humerus, via a ball-and-socket joint.

One of the most important early tasks of these mobile arms was carrying weapons, not for war but for hunting. When this became a male specialisation it followed that males needed stronger arms than females. It followed again from this that male shoulders had to be more massive, with the result that in the shoulder region we see one of the most striking non-reproductive gender differences of the human body. The male shoulders are broader, thicker and heavier than those of the female, a difference exaggerated by the female's wider hips. The typical male body shape tapers inwards as it descends while the typical female shape broadens out.

The shoulders of an average male are about 15 per cent wider than those of the female, but breadth is not the only way in which they differ. Even more important is their measurement from front to back. In this direction the difference is even greater, reflecting the comparative weakness of the female shoulder muscles.

Inevitably this sex difference led to a variety of cultural exploitations. If men wished to appear more masculine, they had only to add some kind of artificial width to their shoulders. The most obvious example of this is the military epaulette, which both stiffens the line of the shoulder and adds to its width with projecting ends. To draw attention to this masculine feature, special badges or emblems of rank are often added to the epaulettes, obliging the eye to dwell longer on the enhanced shoulder shape.

The most exaggerated forms of male shoulder expansion today are found in three very different contexts: Japanese theatre, formal military parades and American football.

In traditional Japanese kabuki theatre the strong, serious, male roles are played by actors wearing vast wings of stiff brocade that almost double their real shoulder width. This garment, known as the *kamishimo*, gives them an immediate air of domination and authority.

The kamishimo began life in the seventeenth century, when it was the official outdoor costume of the Samurai warriors and its quality reflected the status of the wearer. The Samuri were intensely status-conscious and obsessed with personal style, and this may account for the dramatic width and stiffness of the shoulder pieces. The stiffness of the cape-like upper part of the garment was maintained by the use of whalebone and paper.

A strange feature of these shoulder capes is that they make no attempt to conceal the true width of the real shoulders. At the front, the great wings connect to the lower part of the garment by what look like two wide braces. These braces clearly expose the rounded ends of the real shoulders when the wearer is viewed from the front. It is as if the display of artificially super-broadened shoulders is felt to be so powerful that there is no need to go to the trouble of making it more realistic by hiding the real shoulder width. This is an arrogant display that says 'the reaction to wide shoulders is so impossible to resist that a token will do'.

This Japanese fashion lasted, with very little change, from the start of the Edo period at the beginning of the seventeenth century until the early part of the twentieth century. It was only then that this imposingly masculine design began to die out and was gradually relegated to theatrical performances and certain ceremonial occasions.

The military epaulette has a very different design, although the effect is much the same. It creates its impact by adding a stiff, flat strip to the outside of the jacket, along the shoulder line and then ending this strip with a flourish in the form of hanging gold braid. Since very exaggerated shoulders are signs of dominant masculinity, it is not surprising that they were restricted to the higher military ranks.

The name *epaulette* is a French word meaning 'little shoulders', as if a pair of small, additional shoulders is being added to the real shoulders. At different times and in different military organisations, the details of the epaulettes have varied, but always for the same reason, namely the subtle distinctions that have to be made between the different ranks. The precise position of the epaulette, its colour and the length and width of its hanging, bullion fringe were ways of indicating the wearer's rank. Today, the full epaulette is only worn as part of full dress uniform on ceremonial occasions, although in the past it was displayed in the field of battle. Its use on the battlefield by the British Army was abolished in 1855.

American footballers, with their massively padded shoulders, also appear threatening and overpoweringly masculine, even when they are standing inactive at the side of the field. The design of their football pads is amazingly complex. One advertisement, for example, offers shoulder pads with 'high impact plastic body and accessories, air management cushions covered in heavy-duty water resistant nylon and dual density foam in hitting area'.

The puzzling feature of all this high-tech body armour is that other footballers completely ignore it. If it is so essential to the survival of American footballers, then why is it that, in Australian Rules Football, Rugby Union Football and Rugby League Football, the players crash into one another with the same degree of violent contact and yet take to the field with completely unpadded shoulders? The answer, of course, is that the American pads, despite their officially protective function, are in reality masculine display accessories.

Away from the military and sporting arenas, most men today, in ordinary working clothes or business suits, do little to exaggerate the size of their shoulders. Some formal suits have small, inconspicuous pads sewn into the inside of the jacket, to give just a little more weight to the shoulder line, but that is about the limit of it. Anything more would be considered too much of a caricature of masculinity, and as an example of a male trying too hard to appear tough. More subtle measures are called for in ordinary daily life.

When modern women have wished to assert themselves and compete with men on an equal footing, they have sometimes adopted a pseudo-masculine shoulder line, using heavily padded costumes to give the impression of having shoulders as broad as a man's. The emancipated women of the 1890s, the fighting women of the Second World War and the liberated women of the 1980s all displayed square shoulders as a kind of threat display.

There are five gestures involving the shoulders that men employ in different parts of the world. In South America, for example, there is the shoulder brush, in which a man lightly flicks off imaginary dust from one of his shoulders. This is a sarcastic comment, implying that someone is acting in a servile way to gain favour with a dominant individual. The gesture is an acting out of the grovelling subordinate removing specks of dirt from his boss's jacket. It is known locally as *cepillar*, or apple-polishing. The term apple-polishing is taken from the idea of the good boy in the class polishing the apple he is about to give to his teacher to gain favour.

A second shoulder gesture, this one found in Europe and the United States, is the shoulder pat, in which a man pats himself on the back as a joking form of self-congratulation. It is usually performed when he has just succeeded in something, or has guessed correctly in a quiz or a game, and is not being patted on the back by anyone else. It becomes a shoulder gesture simply because he cannot twist his arm far enough to pat himself on the middle of his back and can only reach as far as his shoulder line.

A third gesture is found among the Eskimos in the frozen north. It is the shoulder strike, employed when two men meet. Because of the intense cold, they wear such thick protective clothing that ordinary contact gestures are either difficult to perform or make little impression. As a result the Eskimo males have adopted a vigorous downward striking action on one another's shoulders as a friendly greeting.

A fourth is the Malaysian shoulder clasp, in which the gesturer clasps his shoulders with his own hands, folding his arms across his chest in the process. This is a respectful greeting found in an Asian culture where making contact with another person's body is usually avoided. The man making the gesture is saying: I offer you this hug, and I will do it to myself, rather than intrude on your space.

A fifth shoulder gesture is the shrug, widely used as a disclaimer. The shoulders are raised briefly into a hunched posture and then lowered again, the message being 'I have no idea' or 'I cannot help it', or some other admission of ignorance. By admitting our ignorance we are momentarily lowering our status and, as it sinks, so our shoulders rise.

Shrugging is most commonly found in Mediterranean countries, or in countries with Mediterranean roots. It is rarest in the Orient, especially Japan, where emotional body movements are strongly inhibited by local custom. However, recently, with the intrusion of Western influences, even the Japanese are changing. One elderly male there was recently overheard saying, scornfully, 'Our young men are now beginning to shrug.'

In origin the shrug is a defensive posture. When we are physically attacked or believe we are about to be hit by something or someone, we quickly hunch up our shoulders in a protective movement. The shrug is a token version of this, a fleeting admission that we cannot handle the situation.

If a man receives a shock or a threat that is more serious, he does not shrug but keeps his shoulders up and remains hunched until the threat has passed. For some men, whose adult lives are full of real or imagined threats, there is a tendency to keep the shoulders in a semi-hunched posture much of the time. The down-beaten, down-trodden subordinate who is the focus of hostility and the butt of jokes in his work community or his family life, may easily become permanently hunched and virtually unable to lower his shoulders into a normal resting position. Instead of holding his head high with pride, he holds it low with unavoidable feelings of inadequacy.

In general symbolism the male shoulders usually represent strength and support. We talk of putting our shoulders to the wheel, of walking shoulder to shoulder, of hitting straight from the shoulder, of shouldering a responsibility, or of offering a shoulder to cry on. We carry a hero shoulder high and, when he dies, we carry his coffin in the same way. In some Catholic countries, during local festivals, the men compete to carry the heavy religious statues through the streets on their shoulders. Although today there are many other ways of transporting the huge, holy relics, the shoulders are chosen as the only acceptable form of support. Men who do this on a regular basis end up with one shoulder scarred by a large indentation and permanently lower than the other one, such is their burden.

12. The Arms

The strong arms of the human male played an important role in the early days of the evolution of *Homo sapiens*. The moment when stabbing at a prey with a sharp weapon, the crude method favoured by Neanderthals, was developed into the accurate throwing of weapons, the hunting life of our ancestors took a dramatic leap forward. With their strong right arms hurling spears at their prey from a distance, the improved hunters must have enjoyed a much greater success.

Because it was our male ancestors who became specialised as the hunters in our early hunter/gatherer societies, the difference in strength between male and female arms is considerable. Careful studies have revealed that the average male arm consists of 72 per cent muscle, 15 per cent fat and 13 per cent bone, while the female has only 59 per cent muscle, 29 per cent fat and 12 per cent bone. This difference is reflected in the contrast between male and female javelin records, with males throwing 33 per cent further than females. This is more than three times the difference between male and female speeds in track events.

The power in the male arm comes from the bulging deltoid, biceps and triceps muscles. The deltoid raises the arm sideways, the biceps bend the arm and the triceps extend it. They are anchored to the heavy *humerus* bone in the upper arm. The two bones of the forearm, the more slender *radius* and *ulna*, can be twisted to rotate the hand.

In a recent survey in which women were asked to rate which male muscles appealed to them most, they placed the biceps second only to the abdominal muscles that contract to create the famous 'six-packs' on men who have kept themselves in top condition.

The point where the upper arm meets the lower arm, the elbow, has often been used as a natural weapon. When the arm is bent, the elbow becomes a bony point that can cause considerable damage if used with a swift backward jab. If an attacker approaches from behind, the defensive action of thrusting the bent arm sharply backwards strikes him painfully in the solar plexus. If this fails, a second blow can be made with the arm raised so that the elbow hits upwards and catches him under the chin, or on the side of the jaw. These are recommendations in self-defence classes when an assailant is trying to grab someone from behind and it is surprising just how much injury can be caused using only the elbow.

This type of defence is sometimes seen on the football field, where a player is suffering from tight man-to-man marking, with his opponent literally breathing down his neck and shadowing his every move. A seasoned player waits for the moment when there is an incident elsewhere on the pitch that is diverting the attention of the referee and then lunges backward with his elbow, sometimes causing his marker to crumple up on the ground. By the time the referee has seen the fallen figure, the elbower is already in another part of the field, feigning innocence. Novice players tend to use an elbow-in-the-face in a moment of exasperation, when being shoved and hustled by an opponent in clear view of the referee, and this action is nearly always heavily penalised.

Elbowing has become such a hazard in professional football that it has recently been singled out by the governing body, FIFA, for special attention: 'It has been confirmed by the medical committee to pay attention to a new devil entered in our field of play: elbowing . . . special attention will be brought to the team managers that they shall tell the players not to use their elbows.' Despite this, the attacks continue and in one notorious case a well-placed blow from a raised elbow left an opponent unconscious. He was taken to hospital suffering from concussion, a vivid testimony to the power of this natural weapon on the male body.

A feature of the male arm that deserves closer study than it usually gets is the armpit. Today, most women remove their armpit hair as a way of increasing their femininity. Having less hairy bodies than men means that super-hairless bodies are super-feminine. But for men there is no such urge and the vast majority of them retain their axillary hair-tufts and are happy for them to be seen. Their only concession to modern hygiene is to wash them, usually once a day, and to use anti-perspirant or deodorant sprays before setting off for a social engagement.

This urge to curtail the scent-signalling activity of the armpits is now so massive that the annual worldwide sales of the top six deodorant/anti-perspirant products totalled $8.5 billion at the end of the twentieth century. This anxiety about underarm odours is caused by our modern habit of wearing clothing with sleeves. These sleeves create a tiny closed world of heat and sweat, where fresh, fragrant sweat from the axillary scent glands quickly goes stale and is attacked by bacteria that convert it from an appealing sex signal to an unpleasant body odour.

Originally, in our naked state, the apocrine glands that create the specialised armpit sweat were useful contributors to human sexual foreplay. The scent they produce, that is quite different from ordinary heat-reducing sweat and difficult to detect consciously, was an important arousal device.

The secretions from these glands are different in men and women. Men have fewer apocrine glands, but nevertheless their secretions produce a powerful response in women during close bodily contact. Women who nuzzle close to the freshly washed, naked bodies of their male companions will come under the influence of these primeval olfactory signals, even though they are unaware of the source. A man who has just bathed or showered prior to a sexual encounter and who then, as a last-minute preparation, has sprayed deodorant under his arms, has robbed himself of the assistance of this ancient form of erotic stimulus.

Recent controlled experiments have proved beyond doubt the importance of these male scent glands. The pheromones they produce have been shown to influence the hormonal balance of women. Fresh sweat was collected on pads placed in the armpits of male volunteers. Concentrated extracts from these pads were then placed under the noses of female volunteers for a period of six hours. The women then reported that they felt less tense and more relaxed than they had done before the test. More importantly, there was a significant rise in the female hormone that triggers ovulation. This suggests that if future research can isolate the key chemicals secreted by the male armpit glands, it may be possible to use them as the basis for new fertility drugs or relaxant perfumes for women.

In another test, androstenol, a human pheromone that is chemically similar to testosterone, and which is secreted by the male armpit, was found to produce a stronger reaction in women at the time of the month when they were ovulating.

In studying the way that these armpit scents work, it has emerged that they only operate at very close quarters. The smelling range of pheromones amounts to no more than a few centimetres. In addition to causing pheromones to go stale, thick clothing also blocks them altogether when they are fresh. Even when naked, a loving couple must lie together in such a way that the woman's nose is as close as possible to the man's armpit, and it may not be too fanciful to suggest that this accounts for the fact that women are on average 7 per cent shorter than men. Alternatively, if women are 7 per cent shorter for some other reason, then that would account for why the scent-signalling is focused at the point where the female nose would come to rest.

It has been suggested that genetic differences in male body fragrance may unconsciously play a part in mate selection. In other words, if you are a man, your genes may help to decide whether you smell attractive to a particular woman or not. A careful laboratory study discovered that women respond most strongly to men who have a fragrance similar, but not identical, to their own. Interestingly, they react least to identical or to very different odours. This means that they are most likely to mate with men who are genetically closely related to them, but not too close. This makes sense because it means they will have a bias away from incestuous relationships and also away from relationships with males who are genetically remote. It should be added that women are, of course, quite unaware that they are capable of smelling genetic differences in the men they meet, or that these differences are influencing their choice of male partner. Perhaps that overworked Hollywood cliché, used to describe convincing screen lovers, that the chemistry between them is amazing, has a literal significance after all.

There are considerable racial differences in the odour-signalling abilities of the male armpits. Simplifying, it can be said that, in general, Western armpits have more scent glands than Eastern ones. As a result, Orientals often find Westerners unusually smelly, but are always too polite to say so. In some countries the difference is dramatic. In Korea, for example, half the population have no apocrine glands at all.

As far as we know, the first people to introduce the idea of masking body odour were the ancient Egyptians, who used combinations of citrus oils and spices. They also discovered that the armpit hair provided a large surface area for bacterial growth, and began shaving their armpits. These early refinements applied only to the higher ranks of Egyptian society, of course, and with the collapse of Egyptian civilisation they soon faded into history, with centuries of European stench replacing them. The more fastidious Europeans reacted to the foul odours around them by introducing the art of perfumery to try to obliterate the smells, and this remained the principle weapon until modern hygiene, advanced plumbing and chemical deodorants arrived in the twentieth century.

Today there are several special categories of men who reject the typical masculine display of ruggedly hairy armpits. Muslim men, for example, are taught that they must remove the hair of their armpits by any method that is not too difficult or painful. They are recommended to employ plucking, shaving, cutting it short, using wax treatments, strip hair removers, or special depilatory creams and ointments, and to ensure that the hair should be removed at least once every forty days, or whenever it grows long.

In the West, the male homosexual community and the bondage and sado/masochistic sub-cultures are other groups that tend to favour hairless armpits. Finally, there is that newly recognised category, the metrosexual, where there is also a preference for hair-free underarms. The concept of the metrosexual male was introduced in 1994 when he was defined as an urban male who has a strong aesthetic sense and spends a great amount of time and money on his appearance and lifestyle. An observer of modern male fashions has portrayed him as 'a straight man who styles his hair using three different products, loves clothes and the very act of shopping for them, and describes himself as sensitive and romantic. In other words, he is a man who seems stereotypically gay except when it comes to sexual orientation.'

The replacement of macho man by the new metrosexual man might seem like a gift to the male grooming industry, but there is a catch. The icons of this new trend are footballers like David Beckham and Frank Lampard, both of whom have displayed hairless armpits when applauding their fans, topless, after a shirt exchange at the end of a big match. They, of course, are well known as tough, masculine sportsmen and as famous celebrities. Their ardent heterosexuality is displayed in the gossip magazines for all to see. So, for them to become fastidious and fashion-conscious creates no confusion. But if an unknown heterosexual male were to display overgroomed, narcissistic tendencies, his sexual preferences would automatically be misread by anyone who met him. This limits the metrosexual category largely to famous celebrities who are already publicly recognised for their heterosexuality.

13. THE HANDS

The moment in evolution when we stood up on our hind legs and our front feet became our hands was the time we became truly human. Previously our front feet had been forced to compromise between walking and grasping. Now they could concentrate solely on grasping and become perfected for this single task. With improved, fully opposable thumbs, we could take a firm hold on our environment, both metaphorically and literally.

We developed two kinds of grasping action, the power grip and the precision grip. Male and female hands differed in this respect. Males became much better at the power grip and females at the precision grip. The male hand became stronger and the female hand more flexible. And the difference was considerable, male hands being about twice as strong as those of the female. Even today, with inactive husbands who have never seen the inside of a gymnasium, active wives find they sometimes have to ask their mates to open tight-lidded jars.

Civilisation has not been kind to male hands. Once they were the key elements in tribal success, being the part of the body specialised for fashioning, grasping and throwing weapons, essential for the hunt. In their primeval role they were noble features of the masculine body, but today the term manual labour has a less imposing ring to it. The most successful males now only use their power grips during leisure moments, when holding a tennis racket or swinging a golf club.

Male hands are not only much stronger than female hands, they are also much bigger. This gives male pianists an unfair advantage

144

over female performers, because they have a greater span when the fingers are fully spread. It is also useful for male boxers, and anyone who has ever shaken hands with a heavyweight champion will know how strange it is to feel one's own hand disappearing inside a mass of enveloping flesh. It is said that Muhammad Ali's fists are one and a half times the size of the average male fist.

Although male hands gained a great deal in power during the course of evolution, they did not lose their sensitivity. Anyone watching the speed with which a blind man can run his hands over a page of Braille will appreciate this. And anyone touching a hot stove will know how sensitive to pain these same finger-tips can be, equipped as they are with literally hundreds of thousands of nerve endings.

The underside of the fingers and the palm of the hand is one of the few places on the male body where no hair will ever grow, and the skin will never tan. Even dark-skinned people have pale palms. Another special feature is that the palms, unlike most of the rest of the body, never sweat in relation to excessive heat. They only sweat in response to stress. When we become anxious our palms start to sweat, always a pitfall for a nervous man who is about to shake hands with someone important. There is no escaping the fact that your sticky palms tell the person you are meeting that, despite your relaxed, smiling face, you are really a bag of nerves. No matter how hard you try to wipe them dry, they keep on leaking fear again immediately, and your companion will instantly learn the true effect he is having on you.

Although today it is nothing but a nuisance, this palmar sweating was of great value back in the primal days when anxiety usually heralded a bout of intense physical activity on the hunt. Then it was useful because it lubricated the hands, giving them a better grip. Today, when most anxieties are psychological rather than physical, it is merely an unwanted remnant of the once muscular lifestyle of the human male.

An additional aid to gripping is provided by the tiny papillary ridges on the fingers and palms. These first start to form on the hands of the human embryo when it is about three months old. They start out as tiny bumps, each with a sweat pore at its summit.

At four months these minute volcanoes become connected to one another in mountain ranges, little friction lines that will eventually become the papillary ridges. It is the pattern of these ridges that gives us our fingerprints. And it is because the lines form in a rather haphazard way that each of us has a unique pattern. Even identical twins have slightly different fingerprints. This is because the ways in which the pores fuse into lines are only partially determined by the genes. The fine details of the patterns are determined accidentally by small local variations in pressure on the skin of the embryo's hands as it lies snugly in the womb.

The reason we leave a fingerprint when we hold a glass, or some other smooth, hard surface, is that the tiny sweat pores leak just enough liquid to lubricate the tops of the ridges, even when we may think we are dry-palmed.

Once formed, our fingerprint patterns will stay the same until the day we die, making them the easiest certain way of identifying any individual. Minor cuts or abrasions make no difference. The fingerprint pattern regrows itself exactly as it was before the injury. Only a very deep cut will leave a permanent scar.

On the upper sides of the fingertips are the hard nails that help the hands in two ways. First, they act as local armour, protecting the ends of the fingers. Anyone who has lost a fingernail through injury will know that, before it regrows, any hard impact on the end of the finger is unusually painful. Second, the front edge of the fingernails acts as a gripping device. Presumably in the days before we had nail scissors, or other means of trimming the nails, there was sufficient wear and tear on them to keep them worn down to a convenient level. Failing that they could always be bitten.

There is one strange male custom relating to the length of the fingernails that deserves a mention. Most commonly observed in the Middle East, India and South-East Asia, it is the habit of trimming nine nails short and leaving just one long. The long fingernail is usually on the little finger of the right hand and there have been conflicting explanations of why this is done.

The simplest explanation is that having one long nail demonstrates that you are not engaged in manual labour. It is a sign that you are above such things. The practice was originally Mandarin-

Chinese, but then spread around the Far East. In ancient Japan, holy men, wealthy merchants and noblemen often displayed a long fingernail on at least one little finger. In Thailand it was said to show that you did not work in the paddy fields.

A more specialised function is connected with the drug trade, the long nail acting as a convenient, natural coke spoon for snorting the white power up the nose. It is sometimes referred to as the coke nail. One American school was so alarmed by the prevalence of this use that they included in their ponderous official rules the blunt order: 'The single long fingernail with drug-use implications is strictly forbidden.'

A third explanation given for the wearing of a single long nail is that it is used for sexual purposes. Precisely what these might be is not clear but they presumably include mildly sadistic nippings and stabbings. This may have something to do with another suggestion, namely that the single long nail is the sign of a pimp. One specific sexual explanation was offered by a woman whose new male partner asked her to grow one little-finger nail long and to insert it in his anus when they were having sex, to stimulate his prostate gland.

One ardent nail-watcher has compiled a number of small, practical uses to which he has seen men put their long, little-finger fingernail. These include cleaning the inside of the nostrils; cleaning one's ear; cleaning a child's ear; opening envelopes; scraping chewing gum off the bottom of shoes; opening shrinkwrap; picking up small objects from a flat surface; scratching the head; and defending oneself in prison. Finally, one man insisted bizarrely that he has a two-inch long little-finger nail so that he could paint the Colombian flag on it.

A major cultural difference between men and women has been the extent to which the hands have been decorated. Finger-rings, bracelets, nail polish and henna painting have always been predominantly feminine adornments. These decorations have become so firmly labelled as female that, in general, only the most effeminate of men have enjoyed wearing them. However, there are some important exceptions to this rule.

The most important exception concerns the wearing of a signet ring. This began life as a device for signing important documents. In ancient times official seals were used instead of written signatures and by placing the seal on a finger-ring an element of protection was afforded. A nobleman wore a signet ring bearing his coat of arms, or some other emblem, and he could then press this into soft, heated wax when sealing a document. And loyal followers could swear their allegiance by kissing this ring.

Today, the most famous signet ring is the Ring of the Fisherman, or *Pescatorio*, worn by the Pope. Instead of shaking the Pope's hand when you meet him, the correct form of greeting is to kneel on the left knee and kiss this ring. This creates a problem if you are yourself a great leader and consider yourself to be of equal or higher status than the Pope. Papal diplomacy solves this problem. If the visitor cannot be expected to kneel before the pontiff, the pontiff raises his hand and offers the ring at mouth-level, so that it can be kissed without any subordinate body-lowering. This was done, for example, when Yasser Arafat visited Pope John Paul. Even so, in order to kiss the raised ring, Arafat had to lower his head which, in the photograph of the event, makes it look as though a Muslim leader is bowing down to a Catholic leader.

A new ring is made for each pope and this is placed on the fourth finger of his right hand when he is proclaimed the new pontiff. When he dies, his ring is ceremonially destroyed, being crushed in the presence of cardinals. This is done to prevent the Pope's seal being used after his death, for the backdating or forging of documents. Today this is a mere formality because the ring is no longer used for sealing documents. Since the nineteenth century, the papal seal has been replaced by a stamp that uses red ink.

Many men still wear signet rings today, even though, like the Pope, they no longer use them for sealing documents. However, although the ring has outlived its original function, it still retains its original design, with a broad, flat top, usually decorated with an emblem of some sort, or the initials of the wearer. Even the bling-bling finger-rings of today still retain the same chunky, flat-topped shape as the early signet rings. In this way, a modern male can wear a ring that is purely decorative, but which nevertheless displays a traditionally masculine form.

There are different traditions as to where to place the signet ring. An Englishman, for example, wears it on the little finger of his left hand; a French nobleman puts it on the ring finger of his left hand, and a Swiss puts it on the ring finger of his right hand.

The wearing of a male wedding ring is a much more recent custom, dating only from the twentieth century. It began in wartime, when men who had to leave their wives for a long period of time felt the need to carry with them a reminder of their marriage vows. And it remained as a custom among many men when the war ended, no doubt with a little encouragement from jewellers and wives.

A bizarre superstition arose in Eastern Europe concerning the wearing of male wedding rings. It was believed that, if you wore your ring for more than four hours a day, it would rob you of your sexual potency. As a result, Slavic males would take their rings off periodically to protect themselves from this imagined fate. A suspicious mind might conclude from this that the absence of a wedding ring would make it easier for a man to retain his potency when away from the marital home.

Male bracelets were, until recently, a rarity, although today they are becoming more popular with younger men. Again, a special function often provides an excuse for men to wear this kind of body adornment. This began in ancient Egypt, where high-status males would wear a protective bracelet to guard them against evil spirits.

Today, travellers returning from Africa are often seen wearing an elephant's hair bracelet. The hairs are taken from the tail of the elephant, where they are thick but pliable. Again, wearing this form of adornment is supposed to have protective value, keeping the wearer safe from illness, accident and poverty. According to an old legend, the knots of the bracelet are meant to represent the forces of life, and the strands symbolise the seasons of the year.

One final way in which men can display fancy hand adornments disguised as functional objects is to wear expensive or trendy wrist-watches. These can be diamond-studded and cost a small fortune, or huge and sporty, but in theory they are there to tell the time. Again this gives them a masculine quality and allows a male to be as decorative as his female partner without feeling effeminate.

When it comes to body language, the hands are second only to the face in sending visual signals to our companions. Although women have more expressive faces than men, men are more expressive with their hand gestures. Indeed, in some cultures women are not allowed to make any kind of hand gesture.

There are two kinds of hand gestures. First there are the gesticulations that we perform unconsciously when we are speaking. These help to emphasise our words and they also indicate our mood. They are called baton signals because they beat time to our speech, and the more emotional a statement is the more the hands beat the air.

The mood of the speaker is indicated by the direction of the palms. If they are held palm-up, he is imploring you to agree with him. This is the hand posture of the beggar. If they are held palm-down, he is trying to calm you down, to lower your mood by pressing his hands downwards. If they are held palm-front, he is trying to push you back, to repel or reject you as he continues speaking to you. If his palms are facing one another as he reaches out towards you, he is implying he wants to embrace you with his ideas. If his palms are facing back towards his own chest, he is on the verge of hugging himself or trying to pull you to him.

These baton gestures always accompany speech, but there is another type that replaces speech. They are the symbolic gestures that send a message that could, if we wished, be spoken out loud. Our reason for using them is usually that our companion cannot hear us, because he is too far away, or there is too much noise, or we are under the water, or on the other side of a pane of glass.

Each of the five digits of the hand has its own special significance and its own special gestures. The first digit, the thumb, transmits several important messages. It may point the way, perform the popular thumbs-up gesture meaning that all is well, the thumbs-down meaning that something is no good, or the vertically jerked thumb, signalling a phallic insult.

The use of the thumb as a pointer is less common than the use of the forefinger and in ordinary contexts is considered rather surly. An exception to this rule can be seen at the roadside. In modern times it has become the sign of the hitch-hiker, indicating the direction in which he wishes to travel.

In ancient times it was the downward pointing thumb, aimed at the fallen body of the defeated gladiator, that sealed his fate in the Roman amphitheatre. By jerking the thumb down towards the victim, the crowd could indicate their desire to see him slain. If, on the other hand, they wanted him spared, they covered up their thumbs. This cover-up later became mistranslated as *turned-up* thumbs, giving us the OK thumbs-up sign of recent years. In countries where this mistake did not occur, the more ancient meaning of the upward jerked thumb, which is 'Sit on this!', survives to the present day and causes considerable confusion to travellers and tourists. Hitch-hikers in certain Mediterranean countries have been astonished at the anger caused when they tried to thumb a lift from local drivers, little realising that they were giving them an obscene insult.

The forefinger, or index finger, is the most important of the four fingers. It is the trigger finger, the pointing finger, the dialling finger, the beckoning finger and the finger that presses the button when the bomb goes off. In earlier days, it was believed to be venomous, and it was forbidden to use it when handling medicines.

The male forefinger differs from that of the female in a strange way. In 45 per cent of women, the forefinger is longer than the ring finger, but this is true in only 22 per cent of men. There is no known reason for this difference.

Apart from the widely used gestures of pointing and beckoning, the forefinger is also employed in several obscene signals. Like so many rude hand gestures, these are almost exclusively male. The best known of these is the '*pistola*' in which the stiff forefinger of one hand is pushed through the fingers of the other hand or through a ring made from the curled thumb and the other forefinger. This is such an obvious mime of a penis being inserted in a vagina that it is understood almost anywhere in the world, and in one instance at least has led to the death of the gesturer.

The case of the *pistola* death is a curious one because it involves the creation of the only obscene banknote in history. When the Japanese invaded China just before the Second World War they set up puppet banks in certain Chinese cities. Although these were controlled by the Japanese, Chinese engravers were engaged to make the new banknotes. One of these engravers was so outraged by his task that he added a small detail which at first went unnoticed. The elderly sage depicted on the note, whose hands should have been in a formal posture of reverence, was instead shown making the obscene forefinger gesture. The Japanese authorities eventually tracked down the rebellious engraver and he was publicly decapitated, a high price to pay for the satisfaction of making a rude gesture.

Among male Arabs there is another obscene forefinger gesture that may also bring swift retribution if used unwisely. It looks innocent enough, consisting of no more than the tapping of a forefinger against the bunched tips of the digits of the other hand. In this instance the forefinger is not being used as a phallic symbol but as a symbol of the mother of the person at whom the gesture is aimed. The five digits of the other hand symbolise males with whom the mother has copulated, and the verbal message of the gesture is 'You have five fathers'. There is no record of it ever having led to the death of the gesturer, but it seems likely.

The middle finger is the longest of the digits. It acquired a colourful variety of names in ancient times, the best known being the Roman titles of the Impudicus, the Infamis and the Obscenus. The reason they called it the impudent, the infamous or the obscene finger is that it was the centrepiece of their rudest hand signal, the one that is still popular in the United States two thousand years later, where it is now known simply as the Finger. In this gesture the middle finger is held erect while all the other digits are curled up. This gives it the appearance of an erect penis protruding from a scrotal sack.

The ring finger, the fourth digit, was known to the Romans as the *digitus medicus*, the medical digit, and was used in healing ceremonies. This was probably because, as the least used of the fingers, it was the cleanest. It is also the least independent of the fingers – the most difficult to move on its own. This is why it is also the ring finger, the one on which the bride and groom place their wedding rings, symbolising the fact that they are both giving up their independence.

The little finger, the fifth digit, is often called the pinkie and there are two theories as to how it acquired this name. The first sees it beginning as a Scottish word and a Scottish dictionary published in 1808 records that, at this early date, children in Edinburgh were using pinkie to mean the little finger. It is thought that Scots who emigrated to New York took the word with them and that, from there, it spread around North America and eventually to the rest of the English-speaking world. A second view sees it as a Dutch word *pinkje* that was taken to New York by the early Dutch settlers in the days when New York was known as New Amsterdam.

Among children there is occasionally a moment in play when someone asks for a 'pinkie promise', meaning that their companion must link a little finger with them and then swear an oath. When this is done the promise is said to be binding, and if it is broken the sinister threat is that they must cut off their little finger to atone.

Although this is no more than a childish game, it does have a serious origin. Many years ago, Japanese gamblers called *bakuto* knew that, if they failed to clear a debt, they would have to cut off one of their little fingers as an alternative method of payment. Apart from disfiguring their hand, this also meant that their grip when holding a sword would be weakened and they would be at a disadvantage in any swordfight thereafter.

Later on, the Japanese gangsters known as *yakusa* developed this amputation into a special ritual known as *yubitsume*, which means, literally, finger-shortening. It was a form of punishment, or a way of apologising for some misdeed, or an act that accompanied an expulsion from a *yakusa* group. What made it special was that the man being punished had to cut off his own finger, a much more difficult act than having it cut off by someone else. There were rules for the ritual, starting with the laying down of a small piece of clean cloth. The offender then placed his left hand, palm-down, on the cloth and took up a sharp knife, a *tanto*, and swiftly amputated the end joint of his little finger. To complete the ritual he then wrapped the severed section in the piece of cloth and handed it as his sacrifice to the head of his *yakusa* family, who would have been supervising the event.

Knowledge of this ritual probably explains an otherwise inexplicable act on the part of maverick Hollywood actor Mickey Rourke, who tried to cut off his finger in a fit of anger. He said of the incident: 'I cut my little finger off because I thought I didn't want it. I was angry about something so I decided I didn't need the end of the little finger on my left hand. I didn't cut it off completely, it was still hanging on a tendon . . . It took the surgeon eight hours to sew it back on. I still can't bend it properly.'

In addition to actions that relate to one particular digit, there are many symbolic gestures that involve the whole of the hand. There are more than a hundred of these, many of them with different meanings in different parts of the world. Those that are used by angry men as obscene insults in certain countries can be misunderstood by them if used unwittingly by outsiders, leading to potentially violent encounters.

A sexual gesture used by men in very different ways is the mysterious fig sign. In this gesture, the hand is closed, but with the tip of the thumb sticking out between the bent first and second fingers. In Northern Europe this is a bawdy sexual comment, with the thumb symbolising an inserted penis. The basic message is 'this is what I would like to do'. In much of Southern Europe the meaning changes slightly. There it is more likely to be used as an obscene insult or a threat meaning 'up yours'. In Portugal and Brazil, however, the gesture is used as a defence against the Evil Eye. When other men might 'touch wood', cross their fingers or make the sign of the cross to avoid bad luck, the Portuguese and the Brazilians would make the fig sign. For them it has no sexual meaning, an obvious source of embarrassing errors when travellers move from one country to another.

The use of the fig sign as a protective device may seem odd, but it has its origins in early beliefs that any blatantly sexual display will be so appealing to evil spirits that it will divert their attention and keep them preoccupied. For this reason sexually explicit gargoyles were sometimes placed over church doors, to stop the evil ones from entering the building, and for at least two thousand years little amulets showing a hand in the fig position have been worn by superstitious people, from ancient Rome to modern Rio. Today, however, the wearers are often ignorant of the sexual significance of their lucky charms, and they might be alarmed to know that they are boldly displaying a little carving of an obscene gesture.

The Italian horn sign, in which the forefinger and the little finger are kept stiffly erect while the other two fingers are held bent down by the thumb, has a similar history to the fig sign. To point this horned hand at a man in Sicily could, and in the past probably has, cost the gesturer his life. Known locally as the *cornuta*, the message of this gesture is 'your wife is having sex with someone else', not a comment that is taken lightly around the shores of the Mediterranean. It is unfortunate that students at Texas University proudly employ this horned-hand gesture as their emblem, and use it freely at sporting events, a gesture clash just waiting to happen for American tourists in Italy.

There are several explanations as to why this horned hand should carry such an offensive message. The most popular sees the horned hand as a symbol of the bull. Most bulls are castrated to make them more docile, so the hand sign says 'you have been symbolically castrated by your unfaithful wife'.

This same horned gesture is also used in parts of Italy and nearby regions as a protection against the Evil Eye. When it has this meaning, the horns are usually jabbed directly at the object or person thought to be evil, instead of holding the hand up in the air. In this version the horns belong symbolically to the great, protective bull god of ancient times. Again, there is a two-thousand-year history of this gesture being worn as a small amulet to provide non-stop protection against bad luck.

The hand-ring gesture, with the thumb and forefinger tips joined to create a circle, also has several meanings in different parts of the world. In most places it means OK, everything is fine, but in some parts of the Mediterranean region, in Germany, Russia, the Middle East and parts of South America, it is an obscene gesture that is meant to display an orifice. Today it is usually employed as a sign by a man who thinks another man is a homosexual, or at least effeminate. In parts of France and Belgium it is used to signal that something is disappointing – and here the ring shape is meant to symbolise a zero. In Japan the ring shape symbolises a coin and the message of the gesture is money. In Great Britain there is now a popular version of the ring sign, in which it is jerked up and down mimicking the hand movement of male masturbation.

The V sign is yet another gesture with conflicting meanings. All over the world it is a simple V for Victory sign, with the forefinger and the index finger held out stiffly and separately from one another to form the V shape, while the other fingers are bent. Held aloft like this, there is no doubt about the meaning of the sign, except in one country, Great Britain. There, the gesturer must be careful to perform the hand action in one special way, with the palm of the gesturing hand held away from his body. If he should rotate his hand so that his palm faces his body, and then stab his V sign into the air, his British signal changes dramatically from V for Victory to 'up yours'. This uniquely British insult is rarely understood by foreigners, who sometimes mistake hostility for friendliness.

There is a delightful story that this gesture began at the Battle of Agincourt in 1415, when the French threatened to cut off the bow fingers of the English archers, who were giving them so much trouble in battle. After the French had lost the battle, the English archers are supposed to have taunted them by holding up their two bow fingers, the forefinger and the index finger, to demonstrate that these two digits were still there. Prodding the two fingers in the air, they laughed at their defeated opponents and this is supposed to have given birth to the insulting British V sign. Sadly, hard evidence for this appealing story is not forthcoming. It certainly did not originate at the Battle of Agincourt, because the historian Jean Froissart, who died before that battle took place, recorded English archers waving their fingers at the French during a much earlier siege. But whether this taunting amounted to the making of a proper V sign is unclear. The archer's story also fails to explain the strongly sexual nature of the modern insult. It seems more likely that, for most men who use the V sign today, it is simply a double-strength, phallic-finger insult.

Turning from hand gestures to hand contacts, it is clear that most adult males have a strong resistance to touching or being touched by a stranger, especially another man. Because of this, hand contacts have been formalised over the centuries, giving us one, carefully stylised greeting action that we can perform without feeling embarrassed. It is the handshake, once a local European gesture, but now virtually a global one.

In Europe there had been a long tradition of shaking hands to seal a bargain or bind a contract. That was the original role of the handshake, before it became a greeting gesture. In medieval times it was employed as a pledge of honour or allegiance, and was usually accompanied by a kneeling position on the part of the subordinate. The clasping of the hands was then more important than the shaking element. We do know that the full handshake occurred as early as the sixteenth century because in Shakespeare's *As You Like It* there is the line: 'They shook hands and swore brothers.'

There are now several styles of hand-shaking. There is the politician's handshake in which both hands are used, covering the recipient's hand like a glove. In this amplified handshake, the right hand does the shaking as usual, but at the same time the left hand clasps the other side of the person's hand. It is a favourite gesture of public figures who wish to suggest that they are ultra-friendly, and is sometimes referred to as the glad hand. It is like a miniature hug, with the companion's hand embraced as intimately as possible. The effect is to give a powerful friendship signal while at the same time retaining the formality of this type of greeting. A more intense version has the left hand clasping the companion's forearm, as if edging towards a hug.

One way in which a dominant male can destroy the egalitarian nature of the handshake is to initiate the action by offering his hand in a palm-down posture. This forces the companion to clasp it from below, using a palm-up position. In this way the dominant man can literally gain the upper hand.

There are several rules concerning the etiquette of hand-shaking. The more senior person should initiate the action. A subordinate does not offer his hand first to his superior. A man should not offer his hand to a woman, but should wait to see if she first offers hers to him, unless he is a man of great importance.

There is an odd custom observed by Boy Scouts requiring that they use their left hands for shaking. This is meant to show extra trust because, in theory, the companion could still be holding a weapon in the right hand. Also, it is claimed that the left hand is used because it is closer to a scout's heart. The reality is that employing an unusual handshake gives the action the quality of a secret ceremony. Many secret societies also use idiosyncratic handshakes, with small alterations in finger positions letting the other person know that they belong to a special group or organisation.

An elaborate kind of handshake developed in America in the 1960s was known as the Soul Brother Handshake. It was begun by male African-Americans, but has since spread more widely, and is popular among young adult males as a gesture of close friendship. It is a three-part action, starting with a traditional, palm-to-palm clasp, followed in quick succession by a clasping at the base of the thumbs, and finally, by a hooked clasp of only the fingers. Or this last element may be omitted and the hands return instead to the original, traditional palm-to-palm clasp.

The traditional explanation of the origin of the handshake is that it began in Roman times as a mutual clasping of elbows or forearms, a gesture of trust demonstrating that the strong right arms carried no weapons. Alternatively it has been suggested that it arose as a way of testing the strength of a companion's arm, with each of the shakers feeling one another's arm muscles as a way of assessing their physical power. Today some insecure individuals still employ this power grip, crushing the fingers of the people they meet in the forlorn hope that this will impress them.

It has been claimed that shaking hands is the most common way in which contagious diseases are spread through modern society, and it is not surprising that men generally do not like holding hands with other men for any length of time. There is also the feeling that holding hands is effeminate, and that it might act as a homosexual signal if seen in public. This feeling is not global, however, as there are many countries where adult males do sit, stand or walk together holding hands, as a gesture of friendship without the slightest hint of homosexuality.

This is true of most Middle Eastern countries and also some in Asia and Africa, and there is a famous photograph of an embarrassed President Bush at his Texas ranch walking with a Saudi Arabian prince while holding hands, something the Arab leader found entirely natural but which obviously jarred with the Texan, despite his efforts to keep smiling. As one critic commented at the time, 'some people will do anything to lower oil prices'. Another was even more waspish, remarking that 'in every photo but one,

Dubya is touching His Saudiness with his left hand, his *Arabic ass-wiping hand*!' This is where anti-Bush sentiment overtakes common sense, since if two men walk side by side, one of them must of necessity use the left hand.

A young African-American who visited South Africa reported that there 'It's common for men, young and old, to hold hands. This custom was difficult for me, as an African-American from the land of the gangster pose, to comprehend at first. In my world brothers didn't touch, and those who did were usually confined to the closet.' He concluded by saying, 'It will be a long time before the sight of two African-American men holding hands will do anything other than evoke homophobic slurs or uncomfortable stares. But maybe we can take lessons from overseas . . . In Africa, many of those men holding hands were armed revolutionaries who overthrew the apartheid government. Maybe it's time we all tried a little brotherly tenderness.'

14. The Chest

When our ancient ancestors switched to hunting as a means of survival, new pressures came to bear on the human body. The males who set off on the chase had to develop improved respiration. If they ran out of breath they ran out of food. Compared with other monkeys and apes they had to become big-chested. To house the larger lungs the bone cage formed by the backbone, ribs and sternum had to become more barrel-shaped. It grew in both length and breadth. The male chest became an athlete's chest.

The female developed in a different way. Hampered by pregnancies and infants, she was less mobile. Her chest did not enlarge like that of the male. It developed in another direction, the ribcage remaining small but the breasts swelling to a pair of soft hemispheres. These enlarged breasts had two biological functions, one parental and the other sexual. By contrast, the male chest was simply an improved breathing machine. If it had sex appeal this was secondary, and was because a man with a broader, more muscular chest was a better hunter and therefore a safer prospect as a mate.

Today so many men live unathletic lives that modern male chests are frequently too narrow and too puny to appeal to prospective partners. For individuals who wish to improve their pectoral muscles, so that they can puff out their chests a little more, there is now a simple operation available that offers them 'more defined and sculpted pecs'. A short incision is made in each armpit through which solid, tapered implants are inserted into pockets made by the surgeon underneath the *pectoralis major* muscles. For most men, however, a well-tailored jacket does the trick well enough.

THE CHEST

Once naked, a new problem arises. Male chests come in two varieties, hairy and smooth, and the dilemma is to know which of these two will have the greater erotic appeal. For the man with a hairy chest it is possible, though painful, to depilate this region and display a shining, smooth-skinned chest for all to see. For a man with a hairless chest, the only hope is to wear a chest toupee, an accessory that enjoyed a brief period of popularity in the 1970s, and hope that it does not become detached during moments of intimacy.

When asked whether they preferred a hairy male chest or a smooth one, women were deeply divided in their answers. Some insisted that, because men have generally hairier bodies than women, a super-hairy chest is super-male and therefore very appealing. They added that it not only looks sexy but also feels sexy, like a teddy bear. They also attacked the smooth male chest as being too child-like.

Those women who preferred a smooth chest claimed that lack of hair in that region made men look more youthful and mentioned that, at close quarters, the smooth skin surface was much more sensitive and erotic to the touch. So it seems that the jury is still out on this question.

There may be a cultural variation here, because an Irish singer, Ronan Keating, hoping to make a success of himself in the United States, had his chest waxed to woo American fans. He did not bother to do this physically, but simply had the hairs on his chest airbrushed out of his American publicity photograph. On the cover of his new British single, little tufts of hair could be seen peeping out from his unbuttoned shirt, but in America, on the cover of the same single, these had magically disappeared. This was done, it was said, to make him look less moody and more friendly. Or, to put it another way, less manly and more boyish.

True waxing, on the flesh rather than the photograph, seems to be gaining in popularity in California, and several male Hollywood stars have recently undergone the painful process of having their chest hairs stripped away. The actors Brad Pitt and George Clooney are both reputed to have had this done, reflecting a general shift in a female preference for shiny, smooth male skin.

161

Going against this trend for the sake of his art, Orlando Bloom moved in the opposite direction when he appeared in Ridley Scott's epic film about the Crusades, *The Kingdom of Heaven*. To make him more acceptable as a tough warrior, he was obliged to wear fake chest hair to cover his smooth torso.

It is impossible to leave the subject of the male chest without answering that age-old query: why do men have nipples? This is something that has puzzled anyone who, in an idle moment, has pondered on the question of why an anatomical feature that is functionally concerned with delivering a milk supply to a hungry baby should appear on the chest of the milkless human male. If only a few males had nipples, they would be looked upon as freaks, the male counterpart of the bearded lady. But all males have nipples, so they are part of normal male equipment and their presence requires an explanation.

The way male nipples develop is easy to understand. During the first fourteen weeks in the womb, the human embryo, regardless of its sex, does not display masculine gender features. It is during this phase that nipples can be seen to develop in both sexes. Then, at fourteen weeks, the male hormone testosterone kicks in and masculine features begin to appear on male embryos. But although the new masculine features then go from strength to strength, the nipples are not suppressed, but remain on the chests of both sexes.

This is sometimes given as a sufficient answer to the question 'Why do men have nipples?' but evolution seldom works that way. If a feature is useless it disappears. If nipples are retained, then it is reasonable to suppose that they have some positive value to men, rather than being simply an embryonic fault.

The answer is that they provide an important erogenous zone for the human male during sexual foreplay. According to one report, each male nipple contains around 3,000 to 6,000 ultra-touch-sensitive nerve endings and around 2,000 to 4,000 erogenous nerve endings. The erogenous nerve endings lie just below the touch-sensitive ones and together these two types create a highly sensitive patch of skin.

Surprisingly, it has been discovered that men are either left-nippled or right-nippled when it comes to erotic sensitivity. Just as we all have a dominant left eye or a dominant right eye without knowing about this, so it is with nipples, most men being unaware as to which nipple will give them the greater sexual pleasure during fore-play.

It has been pointed out that Ken, the plastic boyfriend of the infamous Barbie Doll, lacks nipples, and that this has led to some confusion among little girls. Some adults recall drawing nipples on Ken to make him more realistic, and others, who once played with these dolls, offer the absence of Ken's nipples as evidence that male nipples are an erogenous zone and must therefore be banned to avoid inflaming Barbie.

Far from suppressing their nipples, some men actively flaunt them, piercing them and then adorning them with silver or gold rings. Pierced male nipples have existed for at least two thousand years, Roman centurions having adopted the practice to display their courage and virility. In modern times, nipples are just one of the many body parts that are subjected to today's increasingly popular practice of body-piercing. There is even a man who boasts four nipple-rings, because he happens to be one of those rare males who possesses four nipples.

Another key question about male nipples is: do they ever produce milk? Can the human male ever lactate and breast-feed a baby? No less an authority than the great Charles Darwin once contemplated the possibility that the presence of male nipples in humans might indicate that once, long ago, 'male mammals aided the females in nursing their offspring and that afterwards, for some reason, the males ceased to give this aid . . .'. He then reasoned that 'disuse to the organs during maturity would lead to their becoming inactive', however 'at an earlier age these organs would be left unaffected, so that they would be almost equally well developed in the young of both sexes'. Sadly, there is no evidence to support this idea that our direct ancestors displayed regular male lactation, and out of the four thousand species of mammals alive today, there is only one, the Dayak fruit bat, where males have active mammary glands and could assist the females in supplying milk for the young.

However, although it is extremely rare for a man to lactate, it does happen in a few cases. It is most commonly observed where men are being medically treated with female hormones, but it has also been observed, rather strangely, in instances of abnormally high stress combined with poor food. Male survivors of Nazi concentration camps and male prisoners of war returning from conflicts in Korea and Vietnam sometimes demonstrated lactation. Some fathers who have developed abnormally intense urges to breast-feed their babies have managed to stimulate their nipples sufficiently to also produce a small amount of milk, but not enough to provide adequate nutrition.

For many other men, the question is not whether they can produce milk, but whether their chests are too feminine, with swellings that are becoming much too breast-like to suit a rugged male ego. When younger, at school, such males had often been tormented because of their 'man-boobs' and given cruel nicknames like Tits or Booby.

This kind of teasing often leads to deep depression and self-doubt and the males concerned are desperate for some solution. For them, the only hope is cosmetic surgery and 'breast-lifts' are becoming increasingly popular, although still a taboo topic. Last year 14,000 American males went under the knife to have their breasts reduced. True, when compared with the 400,000 women who had their breasts enlarged, this is a small figure, but even so it is still an impressive reminder of how important it is to many males to be able to display a manly chest. As one surgically enhanced male put it: 'Now I don't have to wear a shirt when I have sex.'

When it comes to chest gestures, there are two main symbolic elements involved. The chest region is either used as a representation of the self, or as a sexual zone. The chest as self is seen whenever a speaker wishes to emphasise the concept of 'I' or 'me'. As he uses these words he touches or taps his chest with his fingers. In moments of great happiness a chest-hugging gesture may be performed in which the person concerned embraces his own chest with his arms.

Puffing out the chest or beating it with hands or fists is a masculine display common to many cultures. It signifies self-assertion and it too uses the chest region as if it stood for the whole person. Mourners in ancient times used to bare their chests or beat them in angry grief.

Covering up this 'self' region can act as the opposite signal. In the Orient the crisscross folding of the arms across the chest is a sign of humility that accompanies bowing; and among Arabs, touching the chest, along with the mouth and the forehead, is a polite form of greeting. In Italy, placing the crossed palms flat on the chest acts as a sign of the Christian cross, and is sometimes used when swearing an oath.

Among the sexual chest signals are various forms of breast-cupping with the hands, when men wish to mimic or draw attention to the sexual hemispheres of the female. In Greece, there is a Chest-Thump action in which the male chest is thumped just once with both fists, the fists representing the breasts of a woman.

The simple act of placing a flat hand on one's chest is extremely ancient, going back to classical Greece and beyond. It has been used as a sign of loyalty and also as a way of swearing an oath. For Greek slaves, the left hand held to the chest was a gesture of obedience, signifying that they were awaiting the command of their master.

Today the formal hand on chest is most commonly observed in the United States, where it is employed by non-military individuals during the playing of the national anthem, in place of the usual military salute. On these occasions, the right hand is used, and the origin of the action is obvious enough. It symbolises placing the hand on the heart.

In modern times we think of the heart as the symbol of passions and feelings, but this was not the case in ancient times, when the hand on heart originated. Then, the heart was considered to be the very essence of the person, his intelligence, and the centre of his being. It was this that the ancients were symbolically touching when they stood hand on heart. The brain, in those early days, was seen simply as the instrument of the heart's intelligence. Because of this origin, which still influences us today, placing a hand on someone else's chest is a rather intimate action, usually performed only between lovers or very old friends.

15. THE BELLY

The human belly was accurately described by Dr Johnson as that part of the body that reaches from the breast to the thighs, containing the bowels. In medical terms, it is the abdomen. Although it excludes the genitals, it comes so close to the reproductive zone that it has been subjected to some degree of censorship. Victorians refused to use the word, finding it too vulgar. For them, old-fashioned belly-ache became the more polite stomachache or, for children, tummy ache. Despite the fact that we long ago abandoned Victorian prudery, we still tend to use these inaccurate terms today, prolonging the confusion between the stomach and the belly.

Visually, there are slight sex differences in the belly region. Male bellies are hairier than their female equivalents. On young men at puberty, the abdominal hair starts to show as vertical lines stretching from the pubic region up to the chest. Known in slang as the 'treasure trail' this hair then begins to spread out over the rest of the belly region as the young adult male grows older.

There are four abdominal hair patterns:

The Horizontal. In this the upper limit of the pubic tuft is horizontal and no belly hair grows above it. (Found in 40 per cent of males *under twenty-five.* This is also the pattern found in almost all females.)

The Sagittal. Here there is a narrow band of abdominal hair extending from the pubic tuft vertically to the navel. (Found in 6 per cent of males.)

The Acuminate. This is the typical masculine pattern, with an inverted V of abdominal hair rising from the upper line of the pubic tuft and extending to, or even beyond, the navel. (Found in 55 per cent of males.)

The Quadrangular. Also known as the dispersed pattern, this has abdominal hair scattered more or less evenly all over the belly region. (Found in 19 per cent of males.)

For many men today, the presence of any hairs on the belly is viewed with distaste and there is a growing demand for the waxing of this, and many other areas of the body, to create a smooth, boyish skin, regardless of age.

Apart from hairiness, there are also differences between the sexes in the precise shape of the belly region. In healthy young adults, the male belly is shorter and less curved. More specifically, the distance between the male navel and the male genitals is shorter than its female equivalent. An athletic male has a small, flat, inconspicuous belly that owes its sex appeal to its negative qualities.

One of the dreams of the modern narcissus is to be able to display the infamous six-pack belly, in which the abdominal muscles have been so perfectly tuned that they are rock-hard, can withstand a strong punch, and have the contours of a six-pack of beer cans half buried in the sand. For some males this ideal belly pattern becomes an obsession, fuelled by articles in men's magazines, where it is equated with a condition of masculine physical supremacy that oozes virility and sex appeal.

Special gymnasium devices have been marketed to improve the male belly and create the six-pack contours. They operate on the dubious premise of localised exercise spot-reducing. They promise rock-hard abdominals but often fail to mention that, in order to obtain the desired result, it is necessary to reduce the body's overall fat content from the usual male figure of 12.5 per cent down to about 10 per cent, a change that requires a long-term discipline of carefully designed diet as well as general strength training. Only a truly devoted gymnasium slave can hope for the magical six-pack that produces the idealised belly of youth.

All this changes as age increases. Overindulgence in food and drink sees both male and female putting on weight, but not in exactly the same way. Females tend to run to fat all over their bodies, whereas males show a disproportionate increase in the belly region. There are many older males with spindly bodies supporting huge pot bellies, a silhouette rarely seen in females. It is almost as if the human male has a 'camel's hump' on his belly, where his excess fat store is concentrated, while the female stores her excess fat all over her body.

Back in the days when food was scarce and many poor people went hungry, a large male belly was worn with pride as an ostentatious display of wealth and success. Today, in the more affluent, advanced countries, where the obsessive cult of youth and mounting anxiety about personal health and fitness have come to dominate our thinking, pot bellies are looked upon as sad examples of self-indulgence. Today's politically correct male belly is a flat belly.

The male's pot belly is often referred to as a beer belly, the implication being that regular beer-drinking is a key cause of this condition. Surprisingly, this has been exposed as a myth. Careful research studies have found 'alcohol consumption to be unassociated with weight gain, for reasons not yet understood'. So what is the explanation for the obvious link between boozing and big bellies? The answer lies in the personality of the typical boozer. He is a convivial man who likes to indulge his carnal pleasures, a hedonist who enjoys his big meals, often of junk food, as much as he does his social drinking, and it is the excessive intake of food, rather than the beer, that creates his bulging belly line.

New research in Italy has even discovered that there is a gene for pot belly. If a group of men all overeat and overdrink to the same extent, only some of them will become obese. In an experimental group, those that did develop big bellies were found to share a particular genetic feature in their make-up. Those lacking this feature could indulge themselves without adding to their waistline.

Curiously, these findings were used to give an excuse to pot-bellied males, saying that it was not their fault they were so fat. It was in their genes. Or, to use the official wording: 'Understanding genetic predisposition to weight gain is an essential step in arresting the stigma that obesity is always an individual's fault.' This is a classic example of the distorted thinking of the politically correct who refuse to use the disgusting word 'fat' and resort instead to such phrases as 'horizontally challenged'. The truth is that no healthy person gets fat unless they overeat and underexercise. The genetic factor simply means that, cruelly, some individuals can overeat and underexercise and still not get fat, but that does not excuse the pot-bellied males. If it is not their fault that they eat too much food, then whose is it? When it comes to popping that extra piece of food into the mouth, we are all volunteers.

It should be added that this only applies to individuals who are in good health to start with. For a small minority, with specific glandular problems, especially hyperactive thyroids, it is possible to suffer from obesity without overeating, and they can only solve their problem with special medical treatment.

Some men are brazenly rebelling against the fashionable assault on generously proportioned bellies. As one of them colourfully put it, it is better to die of a heart attack at a younger age after enjoying an uninhibited, happy life, than to linger on as some crabby, miserable, skinny old geriatric vegan. Those who think like him even have their own T-shirt slogan, which reads: 'This Is Not a Beer Belly – It's a Fuel Tank for a Sex Machine.'

This defiant male attitude is particularly common in Scotland, where 43 per cent of men are overweight, with a further 20 per cent categorised as obese. A team of investigators was dismayed to discover that Scottish men would rather be overweight than considered puny. Worryingly, not only did overweight men not want to lose weight, normal weight men actually wanted to gain weight.

Their research was carried out on male shift workers in Edinburgh and Glasgow, who were shown a series of drawings of a male figure in his underpants, starting with a skinny physique with protruding ribs and hip bones, right through to a grossly overweight body shape with a big beer belly. Despite living in a culture full of glamorous images of slim male celebrities, all of the men picked as their ideal body shape one of the pictures showing a clinically overweight male.

This reaction was also confirmed when grossly overweight males were investigated. Health experts were delighted to discover that these men really did want to reduce their bulk, but were alarmed to find that the ideal body shape to which they aspired was not a fashionably slim-line figure, but one that was still technically over-weight.

These findings so horrified the health authorities that they sensibly decided on a change of tactic. Instead of trying to encourage the Scottish male to nibble lettuce leaves and go on a diet, which they accepted was doomed to failure, they urged that instead they should be persuaded to take more exercise.

Exercise without dieting is the golden rule for certain sportsmen on the other side of the world, where big-bellied Japanese sumo wrestlers are looked upon with awe and admiration as major sporting celebrities. They cultivate a gigantic belly for two reasons. It makes them heavier and therefore more able to heave their opponents out of the ring, and it also lowers their centre of gravity, which renders them more difficult to upend. They build up their great bellies by eating a mountain of special stew each day. Called *chanko-nabe*, this is made from fish, poultry, meat, eggs, vegetables, sugar and soy sauce, and is accompanied by twelve large bowls of rice and six pints of beer. Their daily consumption is about seven thousand calories, three or four times that of the average male.

The heaviest sumo wrestler was a Hawaiian called Konishiki, known affectionately as the Dump Truck. At his heaviest, he weighed in at 272 kg (42 stone, or 598 lb) and when he arrived at a hotel the lavatories had to be weight-tested and his bed and chairs rein-forced. Now retired, he weighs six times as much as his wife. Sadly, the average life span of sumo wrestlers is only about forty-five years, but during that period they have the compensation of enjoying more drama and adulation than ordinary men could hope for in ten lifetimes.

How can one explain the modern obsession with flat bellies, one that sees some men spending thousands of pounds on belly-tucks in which a mass of blubber is surgically removed and the then greatly reduced belly is stitched back together again. For such drastic, and presumably painful, surgery to be contemplated means that the desire to avoid a pot-bellied look must be intense. It is argued that a big belly is not only unsightly but is also a danger to health because the deep layers of fat inside the abdominal cavity are close to major arteries and because of this, a fat belly dramatically increases the chance of a heart attack. A big belly also suggests that its owner is self-indulgent, indolent and undisciplined, not qualities that are attractive either to the opposite sex or to potential employers.

The navel is a visual reminder that we were once all babies. If a male navel is surrounded by a soft, rounded belly it looks more baby-like and creates an air of gentle helplessness. By encasing the navel in hard slabs of muscle it is possible to destroy its symbolic vulnerability. When we talk about gut feelings or feeling gutted we are equating the belly with strong emotional reactions. If a man wishes to appear tough by hiding his emotions, he must therefore hide his gut too. By surrounding his belly with hard muscles, he is displaying his ability to control his emotions. If he can control himself, he can control others, and become a successful power player.

In the end it is up to every man to decide which path he wants to take, the gently rounded hedonist, or the lean, mean abdomanic.

In the world of religion there is only one pot-bellied image and that is the Happy Buddha statue, a symbol of good fortune and prosperity. Always portrayed as a bald, portly, squatting figure, the Happy Buddha usually has his naked, rounded belly spilling out of his clothes, and it is said that if you rub his belly it will bring you good luck. This is clearly based on the earlier attitude towards the large male belly, when it symbolised being fortunate enough to have eaten well.

Belly-dancing is always thought of as a purely female activity, but the male belly dancer also has a long history. Back in the days of the Ottoman Empire, which lasted from 1345 to 1922, female belly-dancing began as a form of entertainment in the Sultans' harems and was performed only in the privacy of the palaces. Ordinary men never had the chance to see these exotic dances and the ordinary women of that time would not dream of performing such explicitly sexual dance movements in public.

The solution to the problem was the creation of the male belly dance. The *meyhanes*, the night-time taverns of Istanbul, witnessed the rise of a new form of entertainment, with exotic and outrageous male belly-dancing displays, and men dancing for the pleasure of men. The handsome young belly dancers, called *rakkas*, dressed in sparkling costumes and performed before all-male audiences. Chosen from non-Muslim families, usually Christian, they were highly skilled, having started their dance training at the age of seven. Their training lasted about six years, until they reached their early teens. They would then perform for as long as they could, until their beard started to grow, when they would retire and marry. After that, they would train a group of new boys themselves.

There were two different kinds of male belly dancers, the *tavsan oglans* and the *koceks*. The *tavsan oglans*, or rabbit boys, wore special hats and tight pants. The *koceks* dressed in women's clothing and wore their hair long and flowing. By the middle of the seventeenth century there were estimated to be at least three thousand of these dancing boys, all of them highly skilled, sensuous, effeminate and sexually provocative. Their movements impersonated those of the harem females, with slow belly undulations and suggestive gestures. As a substitute for the forbidden female dancers, these boys became adored celebrities, with enraptured audience members composing romantic poems for them, extolling their beauty. Sometimes they would offer themselves sexually to the highest bidder in the audience.

This tradition of male belly-dancing continued until the nineteenth century, when the male audiences became noticeably more violent. Men would start smashing glasses, arguing, fighting and even killing one another, when competing for the sexual favours of the most handsome boys. Eventually the mayhem became so serious that, in 1856, the Sultan finally banned all male belly-dancing in Turkey. This drove the dancing boys abroad to other countries in the Middle East, where they could continue to perform their exotic dances to appreciative male audiences.

In the twenty-first century, male dancers can still be found in Turkey, but their acts are usually reduced to little more than traditional folkloric displays. However, in sophisticated Istanbul, some nightclubs are now once again putting on the original sexy belly-dance version, with males dressed as females. Interestingly, one of the most famous of these new male dancers refuses to perform for all-male groups and will only appear before mixed audiences, thus reducing any hint of a 'specialist' male sexual display.

The more conservative elements in Turkish society are deeply unhappy about this development and one outraged father even went to the extreme length of chaining his son to his bed for three days in an attempt to put an end to his dancing career. What will happen next will depend upon how mild or how strict Muslim teaching becomes in the Turkey of the future.

The male navel, at the centre of the belly, has aroused far less interest than the female navel. The female navel is often viewed as slightly erotic, because it is a symbolic echo of the female sexual orifice lower down the body. The male navel, lacking this echo, has rarely if ever been viewed as an erotic part of the male body. Its only real interest lies in its religious significance: did Adam have a navel? And if he did, does this mean that God, too, has a navel, since he fashioned Adam in his own image. And if God has a navel, then who was his mother? These are the sort of theological questions that, for many people, reduce religion to a farce. To protect themselves from ridicule, some Muslims have come up with the ingenious idea that, when the first man was created by Allah, the Devil was so angry that he spat on the man's body. His spittle landed on the middle of the man's belly and was about to do so much damage that Allah quickly scooped it off and averted disaster. His action did, however, leave a small scar in the form of the first navel.

Buddhists, who spend a great deal of time contemplating their navels, are engaged in an unusual form of meditation that is not as self-centred as it might seem at first sight. For them, the navel is not a small scar, but the symbolic centre of the universe, so to concentrate on it is to focus, not on oneself, but on the whole of existence.

In complete contrast, for the philosopher Nietzsche 'The belly is the reason why man does not mistake himself for a God.' In other words, the belly is the vulgar, carnal food bag that is diametrically opposed to all things spiritual. This unflattering Western symbolism is at complete odds with Oriental symbolism that sees the belly as the seat of life. In Japan the belly is regarded as the centre of the body and it is for this reason that male ritual suicide there is directed towards it. The Japanese act of *hara-kiri* consists of self-disembowelling with a sharp sword. A literal translation of *hara-kiri* is belly-cut. This is so agonisingly inefficient as a method of suicide that an assistant has to stand by to decapitate the suicide figure, as a way of swiftly putting him out of his misery.

In terms of body language, there are comparatively few belly gestures. We occasionally clasp or embrace our bellies when we feel slightly threatened by our companions. The arms act as a form of barrier signal across the front of our bodies. Unconsciously they say: I must protect my soft underbelly from possible attack. This is a variant form of the more typical body-cross that sees the arms folded firmly across the chest. All such body-barrier movements indicate an uncomfortable social mood, where slight unease is present in personal relationships. Having said this, however, it is important to establish that the person performing the self-hug of his belly has not just eaten a large number of unripe apples, in which case the interpretation of the gesture may have more to do with bellyache than subliminal belly protection.

There are several symbolic belly gestures. The most common is the belly-pat, performed when someone has eaten a large meal, to indicate that they are full up. There are also several local belly gestures that suggest the exact opposite: I am hungry. In Italy, this takes the form of a rhythmic belly-cut in which the flat hand, with the palm down, cuts back and forth sideways across the belly. In Latin America, the fists are pressed hard against the belly, while the mouth is held open, mimicking the agony of hunger pains. And in many countries there is the belly-rub in which the hand clasps the belly and makes a circular motion, implying that the man needs to soothe the pain caused by an empty belly.

This last gesture might be confused with a Central European gesture in which the hand is rubbed up and down on the front of the belly. The message here is 'I am enjoying your misfortune' and is based on the idea that 'I have been laughing at you so hard that you have made my belly ache'.

In France, there is a belly-slice gesture in which the flattened hand, palm up, slices across the belly from left to right. This gesture is associated with the expression '*Plus rien!*' and sends the message that there is 'No more!'

Touching someone else's belly is generally taboo, because of its proximity to the genital region, although two old drinking friends may safely pat each other's beer bellies in a joking way, commenting on their ever-increasing girth. Outside the wrestling arena, the only public context in which a man's belly is likely to come into close contact with another body for any length of time is on the dance floor. The first dance to incorporate belly-to-belly contact was the waltz. Today, this is a dance that seems positively antiquated, but when it first appeared in England in 1812 it was savagely attacked as disgusting and obscene. 'Pollution has entered the ballroom' was the cry. 'The proximity of the partners makes it graceless and vulgar . . . it gives the spectator the idea of Siamese twins of a new genus . . . A fine exhibition truly for the zoological gardens . . . but not for the select ballroom.'

Since the only other time that a young man would press his belly against that of a young woman was during the act of copulation, the frontal body contact of the waltz was simply too much for Victorian eyes. It was described variously as 'will-corrupting', 'calculated to lead to the most licentious consequences . . . and to awaken improper passions'. No wonder it became such a popular craze among the young in the early nineteenth century.

16. THE BACK

Of all the different parts of the male body, the back is probably the most ill-treated. The trouble began millions of years ago when we first stood up on our hind legs and forced our back muscles to work overtime supporting our new, vertical posture. In those days, however, the tribal males were physically active and this kept the muscles in good shape, despite their heavier burden. Today we have become increasingly inactive during our daily lives, and this has meant that our back muscles have grown weak. Sitting all day behind a desk and then going home to relax on soft furniture watching television is not the best way to keep the back muscles, the trapezius, the latissimus dorsi and the gluteus, in tiptop condition. If we then decide to lift a bulky object, or carry a heavy load, we are asking for trouble. Our back muscles, having been out of sight, out of mind for so long, suddenly let us know they are there and that they are not happy about the sudden order they have been asked to obey.

Backache can be anything from a mild annoyance to a crippling agony and it is one of the most common afflictions plaguing modern man. Nine out of every ten men will suffer from it at some time during their lives, most commonly between the ages of thirty-five and fifty-five. And five out of ten will experience it annually throughout their working lives. It ranks as the fifth most common cause of a visit to the doctor. It is the lower back where most of the trouble occurs. It is this region, the lumbar region, that bears the entire weight of the upper body, plus any extra weight that is being carried by the arms or shoulders. Every time we bend, twist or lift something, it is the lumbar region that has to take the strain.

In addition to physical strain, it has been discovered that mental stress, repressed anger and depression can also cause backache. This kind of mentally induced back pain seems to stem from the prolonged muscle tension of certain postures that we hold when we are in an anxious or miserable frame of mind.

Put together, lack of exercise, poor posture and prolonged stress are too much for the male back, brilliantly designed as it may be. It has been calculated that in the United States alone, back pain costs as much as $50 billion each year.

Turning from pain to pleasure, it has to be said that just about the only enjoyment that the back region gives to a man is when he happens to be a masochist who is being whipped. For centuries it has been the back that has had to bear the brunt of serious physical punishment. It seems to have been preferred because it offers a large expanse of skin and because considerable pain can be caused there without damaging vital organs. From the victim's point of view, it is just as well that the skin of the back is thicker than the skin on any other part of the body, and has fewer nerve endings.

Floggings were the traditional punishment in the Royal Navy in earlier centuries, using the cat o' nine tails on the naked back of the victim. That particular kind of whip was based on a Christian theme: Three Times Three, the Trinity of Trinities. The idea was that after being whipped the sailor would be encouraged back on to the path of righteousness. It was known as the cat because, after it had done its work, the sailor's back would look as though it had been clawed by an angry feline. Some sailors cunningly had a Christian cross tattooed on their back in the hope that no ship's captain, however severe, would dare to thrash such a holy image.

Flogging the back was not restricted to the Royal Navy. It was also a punishment meted out in the army and in prisons. In the early penal colonies in Australia a particularly brutal version of the cat o' nine tails was used, having a lead weight attached to each of the nine leather thongs.

Official whippings were abandoned by the British in the nineteenth century, but elsewhere the male back continued to suffer. As recently as the 1990s, some of the Caribbean islands (Antigua,

Barbuda, the Bahamas, Barbados and Trinidad) reinstated the flogging of criminals. In some Muslim countries, even in the present century, strict *shariah* law demands public floggings for a number of crimes.

In Saudi Arabia, Iran, and parts of Nigeria, a Muslim man caught drinking alcohol, gambling or having unmarried sex, may be condemned to a public whipping of 80 or even 100 lashes. In Afghanistan under the Taliban men were even flogged for shaving. In the United Arab Emirates, the penalty for men ignoring traffic regulations is to be flogged in public at the mosque nearest their houses. The sentences are 50 lashes for exceeding the speed limit, and 80 lashes for drunk-driving.

In Iran two businessmen were sentenced to an amazing 339 and 229 lashes respectively for business crimes that amounted to economic sabotage. Also in Iran, a boy who broke his fast during Ramadan was recently given 85 lashes, a punishment so severely administered that it killed him.

In Iraq, when Saddam Hussein was President, it is rumoured that the national football team was given an unusual form of motivation. Saddam's son Uday, who was in charge of national sport, is said to have ordered that, if they did not play well, they should be whipped. Some of the players appear to have experienced this threat as a reality. One of them reported that he was 'flogged until his back was bloody, forcing him to sleep on his stomach in the tiny cell in Al-Radwaniya prison in which he was jailed'. Understandably he later defected to Europe, where the worst form of punishment he could expect to receive for a poor game would be a tongue-lashing.

For those who have managed to avoid backache and public floggings, there are still three ways in which the beleaguered male back can be abused. There was a time when a hairy male back – and some are so hairy that they are almost fur-covered – was considered attractive to the opposite sex, but fashions have changed. Women now seem to prefer their naked men to be shiny smooth rather than assertively shaggy. Enter the wax-stripper. The process is known to devotees as 'manscaping' and it has been suggested that, in addition to making the male back more sexually appealing to women, there is also a covert religious movement afoot to eliminate all traces of affinity with the apes, giving a double incentive for the removal of male body hair.

This creates a dilemma for the modern male; if he waxes his back to a silky smooth sheen, is he being horribly effeminate or, because of the pain involved as the waxed strips are torn off, is he being stoically tough and therefore intensely masculine? The answer to this riddle is still not clear, but by all accounts the popularity of male back-waxing is on the increase, at least in the sophisticated world of trendy urbanites, where young women have been heard to whisper that hairy male backs are gross. For cowards, who cannot face the discomfort of waxing, there is now a special back-hair razor with which a man can safely shave off his own back hair. Known as the razorba, it looks rather like a back-scratcher with a safety razor fixed to the end.

Because of its large, flat expanse, the back is one of the most popular areas for detailed decoration such as tattooing. For the tattooee, it is one of the great ironies of body design that the best canvas is in the worst position, it being difficult for owners to appreciate the display since it can only be studied in photographs or glimpsed in mirrors. There is also another grave disadvantage in having a major tattoo applied to your skin. It is permanent and cannot keep up with new fashions. Back tattoos, some of which are truly complex works of art, have been described as the 'ultimate form of traditional anti-fashion'. They make nonsense of a fashion industry that relies for its economic survival on constant change and the cycling of styles.

Despite this, more and more celebrities are now joining the ranks of the tattooed. Admittedly not all of them are having their backs turned into works of art. Some only venture as far as a small emblem on the forearm, but the fact that they are prepared to visit a tattooist for any sort of needle work indicates a major new social trend.

In earlier days, being tattooed usually identified the wearer as a sailor. Then, in the twentieth century, tattoos were popular with gangsters, bikers, punk musicians and heavy metal band members. Now, in the twenty-first century, it is movie actors, pop stars, famous sportsmen and male models. Tattoos are everywhere, and it is inevitable that fans of the tattooed celebrities will soon be following their idols to the tattoo parlour.

Actors including Robert De Niro, Bruce Willis, Mickey Rourke, Sean Connery, Ewan McGregor, Gerard Depardieu, Colin Farrell, Ben Affleck, Johnny Depp and Nicholas Cage all have tattoos. Nicholas Cage's tattoo design is unusual. He has a monitor lizard wearing a top hat on his upper back. From the world of music, David Bowie, Marilyn Manson, Liam Gallagher, Justin Timberlake, Jon Bon Jovi, Eminem and Robbie Williams also have tattoos. Robbie Williams has the musical notation for 'All you need is love' on his lower back.

Among sportsmen, Diego Maradona, Michael Jordan, Muhammad Ali, Mike Tyson, Dennis Rodman and David Beckham have tattoos. Eccentric American basketball player Dennis Rodman is covered in tattoos, including a Harley-Davidson design, a picture of his daughter, a shark and a cross. Iconic footballer David Beckham has at least nine different tattoo designs. On David's upper back there is a guardian angel and on his lower back the name of his son Brooklyn.

Surprisingly, in the past several heads of state had themselves tattooed, including both Winston Churchill and Franklin D. Roosevelt. Getting tattooed was also a popular pastime for a number of European monarchs, including King Alexander of Yugoslavia, King Alfonso of Spain, King Frederik IX of Denmark, King George II of Greece and Kings Harold II, Richard the Lionheart, Henry IV, Edward VII and George V of England.

It remains to be seen whether the craze for tattooing that is spreading so rapidly today will persist, or if the lasers now used for tattoo removal will become increasingly overworked in the future.

The strange custom of lying on a bed of nails appears to have originated in India where it has been practised for thousands of years by fakirs. These ascetics claim that they wish to demonstrate that voluntary sacrifices are the pathway to enlightenment and to celestial powers. They have been known to meditate lying on a bed of nails for long periods of time, and they believe that if they put their bodies through physical ordeals, they will eventually come to understand the truth.

Some of the self-inflicted tortures to which fakirs subject themselves, such as pushing metal hooks through their flesh, or driving skewers through their cheeks, are indeed horrific, but the fact is that lying on a bed of nails only has the appearance of being horrific. It is something that anyone could do with a little courage and some patience. In recent years in the West it has become a popular magician's trick and, although it sounds potentially damaging, it is in reality comparatively safe.

When a man's back lies on a large numbers of equally spaced nails, its weight is distributed between them so that they do not penetrate the skin. Even when a flat board is placed on top of the man's chest, and objects placed on this board are smashed with a sledgehammer, the nails still do not penetrate the skin of the back. The one problem the man has is rising safely from his bed of nails. This is because, if he does not lift his body with great care, the moment will come when he will only be resting his weight on a few of the nails, and at that second he may suffer a puncture.

Finally, there is an unfortunate gene that surfaces from time to time to create a hump on the back. There is no mystery about what causes the hump – it is a deformity, an extreme curvature of the spine – but there is a mystery about why this particular physical disability should have been considered important in the world of lucky charms. For, in earlier times, it was considered extremely lucky to be able to touch the hump of a male hunchback.

This was especially true in Italy, where it was thought that to rub the hump would provide protection against the Evil Eye. Amulets of a hunchback figure, known by the name of Gobbo, were manufactured in red coral, gold, silver and ivory. They were particularly popular among gamblers who would hold the Gobbo amulet in their hand and rub its hump as the roulette wheel was spinning, the dice were being thrown or the cards were being dealt.

The fame of the Gobbo spread around much of the Mediterranean. It is recorded that small silver charms showing a hunchback were popular amulets on sale in the markets of Constantinople in the nineteenth century. In this same period, they were also the preferred lucky charms in the casinos of Monte Carlo. They even left their mark on the English language. The phrase 'playing a hunch' originally referred

181

to making a play at the gaming tables after touching the hunchback. In France, it was a common practice among Paris stockbrokers to touch the hump of a hunchback before playing the market there.

Even today, in modern Italy, despite growing sensitivity concerning human disabilities, it is possible to buy a plastic keyring showing Gobbo the hunchback as a protection against the Evil Eye. Curiously, Gobbo was always a male hunchback. Female hunchbacks were said to be as unlucky as the males were lucky.

To find a possible explanation as to why a hunchback should be associated with good luck, it is necessary to go back to ancient Egypt where there was a dwarf god of good fortune called Bes. Bes was always portrayed as a grotesque dwarf with a deformed, fat, stunted body, a huge, wide, bearded head and a protruding tongue. He was one of the most popular images employed to ward off evil spirits. His threatening expression, reminiscent of a Maori warrior greeting, and the loud noises he made with the musical instruments he carried, were supposed to frighten off the evil ones. In ancient Egypt his image was everywhere, not only worn on the body but also decorating household goods and buildings. He was employed as a protective spirit by the Egyptians for nearly two thousand years, from 1500 BC to AD 400, and was adopted by both Greece and Rome. It was probably his transfer from Egypt to ancient Rome that started the tradition that led eventually to the Italian Gobbo.

Bes and Gobbo were two kinds of 'little people', a category that also includes gnomes, midgets, leprechauns and elves. Because these 'little people' were so close to the ground, one of their special qualities was supposed to be 'knowing where treasure is buried'. It follows from this that, to touch one of them, might help you to obtain some treasure for yourself. Most of them are untouchable, inhabiting the world of fantasy, but whenever the hunchback gene resurfaces, it creates a person of unusually short stature who is available for touching in the real world. And the obvious place to touch him is on that part of his back that makes him special – his hump.

Attempts to obtain good fortune from a hunchback's hump are still with us today in the twenty-first century. In West African Togo, a church was recently raided because police had been tipped off that there was the amputated hump of a hunchback in a large pot near the altar. When questioned, a church official said that he had bought the hump and other fetishes from a witch-doctor and that it was a magical aid that might attract more people to his church.

Even the world of baseball is not immune to hunchback magic. Back in 1911 the Philadelphia As were so desperate to beat the New York Giants in the World Series that they enlisted the help of a diminutive hunchback named Louie Van Zelst, whose hump they would rub before playing. This gave them such faith in themselves that they did indeed beat the Giants.

17. THE HIPS

The broad human pelvis produces a widening of the body at the point where the trunk meets the legs. We call this projection the hip, deriving the name from the verb 'to hop'. Because the male pelvis is narrower than that of the female, a slim-hipped man is considered to be attractively masculine. The hip width of the average man is 14 inches (36 cm) compared with 15.3 inches (39 cm) for the average woman. This does not seem much of a difference, but the visual effect is magnified by the narrower female waist.

The ideal waist/hip ratio for women is 7:10, that for the male 9:10. When men were shown a series of pictures of women with varying waist/hip ratios and were asked to indicate which was the most attractive, they chose the 7 to 10 ratio. In a similar test, when women were shown a series of pictures of men with varying waist/hip ratios and were asked to indicate which was the most attractive, they chose the 9 to 10 ratio. The response to these contour differences appears to be deep-seated and it is interesting that the distribution of male and female fat deposits appears to support them. When a woman puts on weight, she changes less in the waist region than elsewhere. In this way, although her body may be getting heavier, her crucial waist/hip ratio is to some extent protected. When a man puts on weight, his waistline is not protected in this way.

Wherever there is a slight gender difference in some part of the human body, both sexes are soon hard at work finding ways to increase that difference. So, while women are busy exaggerating their ratio with hip-padding or tight corsets, men are doing their best to display slender snake-hips. The super-male silhouette sees a contour that starts wide at the shoulders and then grows narrower as it reaches the hips. The ideal masculine exaggeration would be to have no pelvic protrusion at all. To this end, male clothing tends to keep the hips as tightly clad as possible. And all movements that involve waggling or swaying the hips have to be avoided, because they emphasise that part of the body. There is one exception to this rule and that is the pelvic thrust. During copulation the jerking movements of the hips that accompany the rhythmic insertion of the penis are quintessentially male.

When Elvis Presley first burst on to the music scene back in the 1950s he was known as Elvis the Pelvis because he gyrated his hips wildly while singing. Young women watching his performances found his movements sexually exciting and not at all feminine. His narrow hips were making rather vigorous, almost violent movements reminiscent of a sexually active male body, rather than the sinuously undulating hips of, say, a Hawaiian dancer. His actions were considered so lewd that television companies were forced to transmit pictures of him only from the waist up. This was done 'in order that younger viewers might not be inflamed by the sight of his hip movements'. Today his gyrations would only raise a smile, but in the 1950s they were viewed as grotesquely obscene.

Frank Sinatra, envious, no doubt, was among those who attacked Presley's performances, calling them 'the most brutal, ugly, desperate, vicious form of expression . . .', typical, he said, 'of every delinquent on the face of the earth'. Outraged clergymen led protest demonstrations at which Presley's rock'n'roll records were publicly smashed. His behaviour was said to be 'a prime factor in the loss of inhibition and youth rebellion'. And all this from a pair of swivelling male hips.

There was a racial element in this anti-hip campaign. Before Presley, white crooners had usually stayed still while they sang, while black performers had moved about more freely. Pentecostal preachers and other religious bigots demanded that radio stations should not play Presley's music, condemning it as 'devil music' and 'nigger music' and calling it sinful, heathen and wicked. Presley himself was labelled as 'that backsliding Pentacostal pup'. A Florida judge threatened to have him arrested if he shook his hips when he was performing on stage in that state in 1956. Presley took his revenge in a subtle way by standing still throughout his perform-ance, but gyrating one of his fingers with mimed hip movements, aiming them at the judge who was in the audience.

The black performer who had epitomised the pelvic approach to dancing, long before Presley's day, was a now largely forgotten nightclub star called Earl Snakehips Tucker. His routines at the Cotton Club in the 1920s and 1930s created a sensation and influ-enced a whole generation of dancers who would later be imitated by Presley to such devastating effect.

The Snake Hips dance was described as 'a contortion, a twisting, shaking type of dance done around the stomach, hips and backside areas . . . taken to extremes'. Duke Ellington, who employed Tucker, said: 'I think he came from one of those primitive lost colonies where they practice pagan rituals and their dancing style evolved from religious seizures.' Tucker, also known as the Human Boa Constrictor, started his act with his body coiled like a snake ready to strike. Then, as his routine progressed, 'his hips described wider and wider circles, until he seemed to be throwing his hips alter-nately out of joint to the melodic accents of the music'. It was this type of dancing, acceptable in the racy atmosphere of a Harlem nightclub in the 1930s, that was to cause so much trouble when it was adopted by a God-fearing young white boy called Elvis in the austerity of the post-war 1950s.

In the twenty-first century, Tucker's and Presley's legacy is to be found in hip-hop and freak dancing styles that are once more causing parental outrage because of the explicitly sexual nature of the hip movements involved. Recently, some high schools in America have demanded that both students and their parents sign a form prohibiting 'intimate touching, sexual squatting or sexual bending' during school dances. They are also banning freak dancing that involves 'grinding in sexually explicit positions, typically with the girl's back to the boy's front'. The potency of human hip move-ments has surfaced yet again.

In some countries, especially in South America and the Middle East, there is a popular obscene gesture, the hips-jerk. This has nothing to do with dancing and is used by a man simply as a sexual signal saying, 'This is what I would like to do with her', or 'This is what they are going to do'. It is the silently mimed act of male pelvic-thrusting. The man, who is standing, thrusts his hips forward rhythmically while his elbows are held to his sides. In the South American version, the forearms are bent forward and are jerked backwards each time the hips move forward, miming the act of holding on to the female body while copulating. In the Middle Eastern version, which appears to have arisen independently, the hips jerk back and forth while the arms are held still.

As far as hand gestures are concerned, there is only one important male hip action and that is the arms akimbo, in which a hand is placed on each hip, with the elbows pointing outwards. This is an unconscious action we perform whenever we feel anti-social in a social setting. It is observed when sportsmen have just lost a vital point, game or contest. It is as if they are automatically adopting an 'anti-embrace' posture without recognising what they are doing. It also occurs when a few men are standing together and one of them wishes to exclude someone else from their small group. Without thinking, he will raise a single arm into the akimbo position, aimed at the man who is to be kept at a distance.

The akimbo posture is saying, in a sense, 'I don't want to be embraced by anyone at this particular moment, so please keep away'. It is almost as if the elbows, sticking out at the sides of the trunk, are giant arrowheads pointing outwards. Or, perhaps, archers' bows ready to fire arrows out sideways. This may explain the origin of this strange expression. Centuries ago, when archers were still active in battle, the word akimbo was written 'a ken bow', meaning 'a keen bow', which can be interpreted as a 'sharp-pointed bow', as distinct from a curved bow. The sharp point, of course, is formed by the outward pointing elbow. Strangely, the word has no equivalent in other languages. It is sometimes given a descriptive phrase such as 'fists on haunches' or ' the pot with two handles', but there is no single-word equivalent, despite the fact that the action is used globally. This reflects the fact that we perform the gesture without thinking.

There is one profession where the akimbo action is specifically used as a mild insult, and that is among stage actors. After a performance, if one of the cast has been overacting, another can be heard to say 'he was a bit akimbo tonight', meaning that he was using too much body language.

In South-East Asia, especially Malaysia and the Philippines, the akimbo action is adopted as a specific signal of seething rage. This is merely an exaggeration of its ordinary use, taking the 'upset' feelings of the usual akimbo posture and extending them into full outrage or anger.

There is also a slightly modified version of the hands-on-hips posture, in which the elbows still stick out sideways, but the hands are brought forward and the thumbs are hooked into frontal pockets or a belt, with the fingers visible and pointing towards the concealed male genitals. Here, the emphasis is shifted from the outward pointing of the elbows to the inward and downward pointing of the hands, in the direction of the penis. Not surprisingly, this is a popular gesture among sexually assertive males who, as one young woman complained, are unconsciously 'willing you to look, touch and admire the part they are proudest of'.

When we are young adults, the power of our hip thrust is an essential part of our athletic ability. As one trainer put it: 'Elite athletes in all sports have one thing in common . . . strong and explosive hips. Development of core muscle-strength and explosive strength are essential to excelling in sports.' In old age, however, the hips tend to let us down. The ball-and-socket joint gets badly worn and, among the elderly, walking sticks, Zimmer frames and eventually wheelchairs are increasingly in evidence.

As upright walking apes, it seems we still have quite a way to go in perfecting our novel mode of locomotion. The problem, of course, is that we will already have bred and passed on our genes long before this hip decline takes hold, so there will be little evolutionary pressure to help us. Modern surgery, however, is always improving and today hip replacement surgery is becoming increasingly efficient and more common. For some unknown reason, the human male is far better off in this respect than the human female. Hospitals report that the ratio of female to male hip replacements is 4:1.

Finally, a word about the terms hipster, hip and hippie. One could be forgiven for thinking that these have something to do with the hips of the male body, but they do not. The term hipster was used in the jazz and swing cultures of the 1940s and 1950s, where it meant that someone was in the know, or hip. The word hip in this context comes from turn-of-the-century military slang. When the drill sergeant called out 'hip-two-three-four', a corruption of 'up-two-three-four', the group that was well synchronised was said to be hip. Those in the group knew what they were doing because they all brought their legs hip (up) together. In the 1960s, a new type of youth culture emerged, the long-haired, drug-taking, love-promoting social rebels of that lively decade. They borrowed the word hipster and called themselves hippies. They would no doubt have been horrified to discover that, in addition to being in the know, their stridently anti-military movement had as its chosen title a term relating to military efficiency.

In the twenty-first century, hipster culture has resurfaced but in a new form. The term hipster now refers to those who are 'devoted to ironic retro fashions, independent music and film, and other forms of expression outside the mainstream'. There is also a new middle-class sub-culture calling itself the neo-hippies, the modern equivalent of the long-haired 1960s hippies. They are anti-capitalist and anti-chav, with their emphasis on recent causes such as animal rights, gay rights, women's rights, organic produce, recycling, breast-feeding and environmentalism.

18. THE PUBIC HAIR

Pubic hair first appears as a triangular patch of short curly hairs, just above the penis, as boys reach puberty, usually in their thirteenth year. This, of course, is when the sex hormones come into operation and, in addition to initiating sperm production, they stimulate the sudden growth of body hair in the male. As testosterone levels increase, the sequence of appearance of new hair reflects the hormonal sensitivity of the different regions. The pubic area is most sensitive, and the new hair usually grows there first.

In boys, the very first pubic hairs appear sparsely on the scrotum or at the base of the penis. A year later, the hairs around the base of the penis are too many to count, and within three to four years, hairs fill the pubic area. Unlike the male's head hair, his pubic patch never goes bald and never goes grey, as he grows older. So, pubically speaking at least, he remains forever young.

This distinctive patch of pubic hair acts as an immediate visual signal indicating sexual maturity and, in the days when we went naked, could be detected even from a distance. It has been suggested that this was its primary function during the early days of the human story, at a time when most of the rest of the ancestral coat of fur had been lost.

It is certainly true that in pale-skinned races the dark triangle of hair is highly conspicuous. The argument is less convincing for dark-skinned people, however, and it seems likely that it had other functions as well. Two have been suggested. One is that the tufts of hair acted as buffers to prevent skin chafing during the vigorous and often prolonged pelvic-thrusting of face-to-face copulation.

190

Bearing in mind that whole cultures have practised depilation of their pubic regions without any recorded skin damage, this too seems a little far-fetched.

It is the third function that appears to be nearer the truth. The suggestion here is that, as with the tufts of hair in the armpits, the pubic tufts are essentially scent-carriers. There is a powerful concentration of apocrine scent glands in the crotch region and dense hair there acts as a scent trap for the pheromones they secrete. As with other parts of the body, tight clothing can easily create problems for this scent-signalling system, allowing the secretions to go stale and converting the natural sexual fragrances into unpleasant body odours, but in the naked, primeval days, when dramatic evolutionary changes were taking place in our species, it would have worked well enough.

At puberty, boys are always proud of their newly sprouting body hairs, but some girls are less enthusiastic. They are fully aware that their pubic hair is a badge of sexual maturation, but unconsciously they may feel that it is somehow masculine because the bodies of adult males are generally much hairier than those of women. As a reflection of this there is a quite irrational fear of hairy spiders among females at the age of puberty. Among ten-year-olds there is no difference in spider reactions between boys and girls, but by fourteen spider hatred is twice as strong in girls as in boys, and the word hairy is always added descriptively with a shudder.

Some pious religious writers have suggested, in all seriousness, that the true function of the pubic hair is to mask the disgusting genital details, and early writers who dared to discuss such matters were of the opinion that this was a wise strategy on the part of the Almighty. However, if God were attempting a cover-up, he failed dramatically with the human male. Both the penis and the testicles, although heavily fringed with hair, remain stubbornly conspicuous.

The Deity would undoubtedly have been pleased to learn that in Korea a lack of female pubic hair is sometimes a cause for serious concern among amorous males and that some brave women there have gone as far as having hair surgically transplanted from their head to the genital area to attain the desired amount. Unfortunately their primary motivation does not appear to be to create a genital cover, but rather to satisfy their males' erotic craving.

This drastic measure is strangely out of step with current pubic trends, for there is a widespread tendency at the present time for the depilation of the pubic region in both females and males. There are five reasons for this. Firstly, there is a matter of hygiene. Naked genitals are potentially cleaner than hairy ones and any disease that might possibly be lurking in the dense hairy jungle has a slim chance of surviving on a naked skin surface. In particular this includes the notorious 'crabs', or pubic lice, that like to cling tightly to the genital hairs and suck the blood locally when hunger takes them.

Secondly, there is the increased vulnerability and exposure of the external genitals, with their precise shape more clearly delineated. Thirdly, there is the juvenile look of the hairless pubic region, a look that enables an adult male to shed some years, and that will inevitably make him more sexually appealing in a culture increasingly obsessed with youth.

Fourthly, some males have noticed that removing the pubic hair makes the penis appear longer. And finally, it has been pointed out by some women that a genitally hairless male does not shed telltale, incriminating pubic hairs during illicit encounters.

Annoyingly, pubic hairs do seem to become detached rather easily. For those who relish bizarre statistics, it has been calculated that a man living an average life span of seventy-five years will shed a total of 45,260 pubic hairs. To express it another way, each year the male population of the world will shed a total of 2,190 billion pubic hairs. And with the new reliance on DNA testing, quite a few of these will no doubt end up in court.

There are, therefore, good reasons to remove pubic hair, but among the male population of the world who are the ones most likely to take this step? A major group are the Muslims, guided by Mohammed, who is reported to have said: 'The natural state of man is five things: circumcision, trimming the moustache, cutting the nails, plucking the armpit hairs and shaving the pubic hairs.'

In the Muslim faith it is a required custom to remove the pubic hair and hair under the armpits for the purposes of general hygiene and cleanliness. Hair in such areas is considered unwanted. It is strongly discouraged, but not forbidden, to let them grow for more than forty days.

For Muslims the removal of pubic hair is essentially an act of personal grooming and has no particular sexual connotation. For other males, it has become a modern fashion, following in the wake of female hair-waxing, and for them it is not only a matter of cleanliness but also has a strong sexual element. This is because they are treating the most sexual part of the body as a focus of attention. By dwelling on the pubic area and carefully transforming it, they are automatically increasing the erotic value they place on their bodies. Whatever the specific advantages may or may not be, there is a general heightening of sexual alertness that accompanies such modifications, and that helps to keep this new fashion alive, even if it is tedious to maintain. But of course fashions never last forever, and the cry that a hairy-bodied man is a thoughtless primitive and now outmoded may well one day be replaced by a return to favour of the natural man, who sees all primping and grooming as unduly self-conscious and narcissistic.

Finally, for those men who are already rejecting the naked body trend, an advertisement recently appeared offering Pubic Hair Beads for Men. It has been observed that the clitoris, the most sensitive erogenous zone on the female body, is not directly involved in the friction caused by the pelvic thrusting of human copulation. In order to correct this weakness in the system, it has been proposed that men should attach special, smooth beads to their tangle of pubic hairs, in such a way that their surfaces would rub rhythmically against the clitoris during penetration. The advertisement for these beads reads: 'We are proud to introduce Pubic Hair Beads for the caring male . . . thus improving stimulation for his partner. Available in burnished bronze, ironwood, petrified bone and walrus tusk ivory.'

19. THE PENIS

Compared with the penises of other apes, the human penis is most unusual. It is much longer, much thicker, possesses a strangely shaped tip, and lacks the penis bone that aids erection. Being such an oddity, it is hardly surprising that it has been the focus of so much interest, both academic and pornographic. Dozens of books have been written about it, taboos have arisen over it, laws have been enforced about it, and it has been the subject of a million jokes.

Even in our modern, permissive society, the erect penis remains one of the most forbidden of all images. Something as intrusive and intimate as open-heart surgery can be screened on television, as can pictures of the wounded and mutilated bodies following a bomb outrage, but the human phallus is still outlawed as obscene. As someone once summed it up: 'You can show a gun, that shoots death, but not a penis, that shoots life.'

For the purposes of this book, this taboo must be swept aside and, like all the other parts of the male body, the penis must be examined objectively. First, its basic anatomy.

The average human penis, when fully erect, has a length of 6 inches (15.2 cm) and a circumference of 5 inches (12.7 cm). These figures are based on the famous Kinsey study of 3,500 American males. Later investigations produced much the same result; one, of 3,000 men, gave the average figures as 6.3 and 5.1 inches (16 and 13 cm); and another, conducted by a contraceptive company, also on about 3,000 men, from 27 different countries, gave 6.4 and 5.2 inches (16.3 and 13.3 cm).

194

The length of the penis when not erect is 3 to 4 inches (7.6 to 10.1 cm) and its circumference is about the same, so the enlargement during erection is considerable. It is achieved by a remarkable mechanism, in which the local blood circulation is disrupted, creating a kind of 'traffic jam' in the blood inside the penis. The vessels inside the penis are dilated, while the draining vessels are compressed. While this is happening, the penis not only enlarges but also becomes much stiffer and, instead of hanging down, rises above the horizontal. In 20 per cent of men it rises to forty-five degrees above the horizontal, and in 10 per cent it reaches a vertical position. In uncircumcised men, this action pulls back the foreskin and exposes the sensitive glans at the tip of the penis.

When inserted into the vagina in a typical sexual encounter, the penis will make between 100 and 500 thrusts before ejaculation is reached. This is so much more than other primates. A typical monkey ejaculates and dismounts after only a few thrusts; for example: 2–8 thrusts in the macaque monkey; 3–4 in the owl monkey; 5–20 in the howler monkey.

Three variations in erect penis shape have been identified: the Blunt Type, the Bottle Type and the Prow Type. The Blunt Type is the most common, with a straight shaft and a slightly wider head. The Bottle Type has a shaft that is slightly wider than the head. And the Prow Type curves slightly upwards towards the tip, giving the penis a banana shape.

The Prow Type is of special interest as it appears to be a male adaptation to the position of the female G-spot. This small erogenous zone lies on the frontal or upper wall of the vagina and extra pressure upon it during pelvic thrusting increases female arousal. By curving upwards, the Prow Type penis will automatically increase this pressure, as the erect penis moves up and down the vaginal tube. In evolutionary terms, this must be seen as the most advanced form of human penis.

The *length* of the penis ensures that, when sexual climax is reached and ejaculation occurs, the seminal fluid containing the sperm will be deposited at the far end of the vaginal passage, close to the cervical opening. This dramatically increases the likelihood of sperm successfully migrating up towards the egg.

The *shape* of the penis ensures that any seminal fluid already present will be displaced. It has been argued that the glans at the tip of the human penis has evolved its unusual design as part of a strategy that will help to defeat female infidelity. If a man's female partner has been unfaithful and is carrying another male's seminal fluid inside her vagina, it is a great advantage for the male partner to own a penis with an enlarged head that will act as a plunger, squeezing out the unwanted semen. The glans, sitting like a helmet on the end of the penis, has a protruding rim called the corona or coronal ridge. As the penis plunges deep into the vagina, any liquid already close to the cervix will be squeezed back beyond this rim and then, as the penis is withdrawn, it will be expelled and lost. Then, when the male partner himself ejaculates, his seminal fluid will successfully replace the rival male's fluid. Significantly, at the very moment of ejaculation, the human male feels a sudden urge to stop thrusting. This is important because, if he did not stop, he would be in danger of displacing his own seminal fluid.

To test this semen displacement theory an American research team simulated sexual encounters using dildos and artificial models, and they found that deep thrusting of the artificial penises did indeed displace the semen, with great efficiency. Short artificial penises did not have this effect, nor did long ones that were not thrust deeply enough. Interestingly, a questionnaire given to six hundred young adults revealed that pair-bonded males tended to thrust deeper in circumstances where there was an increased likelihood of female infidelity. In cases where a male partner had explicitly accused his female of cheating, the thrusting was even more vigorous.

This theory provides an explanation of the unusually large dimensions of the human penis and its unusual shape, but there is an alternative suggestion that deserves consideration. This sees the special design of the human penis primarily as an arousal device. It was mentioned earlier that the greatly increased parental burden of the human species led to a reproductive system based on pair-formation. With a serial litter to care for, the human female needed the help and protection of a long-term male partner. His presence would double the parental care and greatly increase the survival chances of their offspring.

Intense mutual pleasure gained from extended sexual encounters would help to cement this human pair-bond. Pelvic-thrusting in a typical monkey lasts about eight seconds. In human beings it lasts, on average, eight minutes, or sixty times as long. And it may last much longer, even up to an hour in some instances. In addition, the increased width and length of the human penis, stretching and filling the vagina, would greatly increase female arousal. The design of the glans, with its fleshy ridge protruding upwards but not downwards, would provide additional stimulation to the upper or frontal wall of the vagina, which is precisely where the highly sensitive G-spot is located. Viewed in this way, the human penis becomes not so much a plunger as a pleasure device, arousing the human female to an orgasmic level unknown in other primate species.

The feeling of pleasant exhaustion following orgasm usually results in the female remaining in a horizontal position for a while afterwards, with little body movement. This is another important aspect of the human sexual encounter, since a quickly resumed vertical posture could cause considerable semen loss. It is probably no accident that the only bipedal primate is also the only one with an intensely orgasmic female.

To sum up, the large, uniquely shaped human penis and the uniquely intense female orgasm together create a reproductive system in which foreign sperm are displaced and the partner's sperm are retained. Also, the greatly increased and prolonged sexual arousal of both partners helps to cement and protect the human pair-bond. Whether this system is primarily concerned with pair-bonding, and has a secondary impact on sperm control, or whether sperm displacement and retention is primary and pairing is secondary, is not clear at present. There is, in any event, no conflict between the two theories; it is merely a matter of evolutionary priority.

Turning to the mechanism of ejaculation itself, it consists of two phases. The first involves the movement of the seminal liquid and its contents, from its various sources, into a position ready for emission. When this happens, the male experiences a sensation that ejaculation is imminent and inevitable. There then follows a pause of two to three seconds, during which no conscious control can stop or delay the process. At the start of the second phase, there are strong expulsive contractions of the penis, propelling the seminal fluid from its tip. The first two or three of these contractions are so powerful that, if the penis is outside the vagina, the seminal fluid may be expelled as far as 24 inches (61 cm).

When the contractions have died down, the erect penis returns to its flaccid state and at this point rearousal is impossible. The length of time it takes for an erection to be regained varies from male to male and, most markedly, with age. The quickest recovery period observed by Masters and Johnson was with one young male who was able to ejaculate three times in ten minutes, but this was exceptional. After the age of thirty, most males find it impossible to achieve a second ejaculation after only a short rest. Eventually, as ageing proceeds, the male can only muster a single ejaculation and must then wait for a day or more before his next sexual climax. In elderly males, the explosive ejaculation of youth becomes little more than a seminal seepage. This is connected with the fact that levels of testosterone at the age of seventy-five have fallen to half what they were at twenty-five. Furthermore, between 70 and 80 per cent of men in their seventies are incapable of attaining a full erection. In evolutionary terms this makes sense because, at that advanced age, so close to senility or death, they are ill-equipped to care for any new offspring they might father.

In the world of sports and athletics there is a strongly held belief that ejaculation of sperm somehow weakens the male's physical condition, and should be avoided when preparing for a contest. Many trainers and coaches insist on sexual abstinence before a major event, but Masters and Johnson, in their detailed study of human sexuality, state bluntly that there is no physiologic evidence to support this theory. If it does work, it must be because it makes the men involved so angry that their urge to defeat their opponents is heightened.

Recent findings by American research workers have suggested that there is something special about the seminal fluid in which the active sperm are floating. In addition to being the liquid that helps to carry the sperm up to their destination, it is claimed that it also has a direct chemical impact on the female body. When sexually active young females were studied, it was found that those whose male partners did not wear condoms gained a hitherto unsuspected benefit. It seems that the semen in their vaginas influenced their psychological state in some way, making them generally happier than those females whose partners always wore condoms.

The controversial conclusion was that mood-improving hormones contained in the semen were being absorbed through the vaginal walls. This is certainly possible and would make good evolutionary sense, creating a sense of wellbeing from the indulgence in potentially procreative activities, and therefore increasing the desire to reproduce. There is, however, a flaw in this conclusion, because the women concerned were well aware of whether their partners were wearing condoms or not. It could be that those who were having protected sex were unconsciously unhappy about the thought that there was no chance of procreation. This does not mean that they were secretly hoping for a pregnancy, but rather that, at a deeper level, they were responding less intensely to recreational sex than to procreative sex.

An interesting question arises from this research, namely, do a couple feel more intensely aroused by safe sex using a contraceptive pill, than safe sex with a condom? It is true that in both cases they are aware that they are engaging in recreational sex, but in the case of the pill there is no direct evidence of this. They cannot feel the pill in the way that they can physically feel the presence of the condom, and this may mean that, psychologically, sex with the pill is more fulfilling.

Cultural attitudes towards the penis have varied greatly. In Western society the general approach has been to keep it hidden and, in polite society, to pretend that it hardly exists. There was, however, one extraordinary exception to this rule: the codpiece.

The codpiece began as a modesty covering for the male genitals. Mens' fashions in the fourteenth century had changed so that, with a shorter doublet, it was thought necessary to cover the crotch region with a special pouch. The word 'cod' in this context has nothing to do with the fish of that name, but comes from the Old English, meaning a bag or scrotum. So the codpiece was a scrotum-holder, introduced to ensure that there was no accidental exposure of men's 'privvies'.

Once it was in place, however, it gradually became more conspicuous and prominent, until, far from being a modesty covering, it
became a blatant advertisement of male sexuality. Padded, decorated,
ornamented with jewelled pins, and lovingly shaped, it rose from
between a fashionable man's legs like the prow of a proud ship.
Even suits of armour displayed carefully sculpted codpieces. As time
went on it became crescent-shaped and eventually rose vertically,
creating the impression of a protective covering for a permanent
male erection.

The sixteenth century French satirist François Rabelais lampooned
the extreme form of the codpiece, describing one in the following
words:

> It was fashioned on the top like unto a Triumphant Arch, most
> gallantly fastened with two enamell'd Clasps, in each of which
> was set a great Emerauld, as big as an Orange; for . . . it hath
> an erective vertue and comfortative of the natural member.
> The exiture, out-jecting or out-standing of his Codpeece, was
> of the length of a yard, jagged and pinked, and withal bagging,
> and strouting out with the blew damask lining, after the manner
> of his breeches . . .

For some men, however, the enlarged codpiece had a far from
erotic function. Members of the nobility who were suffering from
syphilis employed it as a protective covering for their medication
and special genital bandaging. For them, it was a blessing to be
able to disguise their painful, sad condition as a rakishly phallic
fashion statement.

Needless to say, the Church was horrified by this 'sinful clothing',
but nothing could stop it, until eventually, as happens to all costume
styles, it fell out of fashion. This happened in the sixteenth century
during the reign of Elizabeth I, and from her day until the era of
Elizabeth II it has never been seen again as a major fashion. In
modern times, it has only resurfaced on exotic costumes worn by
some of the more eccentric pop stars, or has appeared briefly in
sci-fi movies.

In the 1970s there was one abortive attempt to bring back the codpiece as a standard item of male clothing, but it was too explicit even for the most rebellious of young males. In 1975, Eldridge Cleaver, an American civil rights leader and member of the Black Panther Party, tried to introduce a new line of men's tight-fitting trousers featuring what he called a Cleaver Sleeve. This was essentially a large sock that protruded from the front of the trousers, accommodating the penis and allowing it 'free movement and size changes'. Bizarrely, Cleaver claimed that his new fashion design 'would revolutionize sexual attitudes in a way that would ultimately eliminate such crimes as rape. They would also abolish . . . the crime of "indecent exposure" and replace it with "decent exposure".'

Despite Cleaver's campaign, no males in ordinary everyday clothing have dared to reintroduce the codpiece as a masculine fashion accessory, although it may yet enjoy a renaissance if our increasingly liberal attitude towards explicit sexual display continues, and future fashion designers find themselves running out of ideas.

Looking further back, to ancient civilisations, there have also been times when the penis has been celebrated rather more boldly than it is today. The ancient Greeks, for example, regarded the foreskin as especially beautiful and looked down their cultured noses at their neighbours, such as the Phoenicians, Egyptians, Ethiopians and Syrians, who practised foreskin removal. The Greeks rejected the religious excuses given for their 'blood rites of penile reduction', as they called them, and even managed to persuade the Phoenicians to abandon their genital mutilations.

The Greek obsession with the foreskin was such that they even had special names for different parts of it. The section covering the glans was called the *posthe* and the tapered part that extended beyond the glans, ending in the orifice at the very tip, was called the *akroposthion*. It was this terminal part that seemed to be the main focus of their attention, and the longer it was the better they liked it.

During gymnastics and when competing in the Olympic Games, Greek athletes were completely naked. (The presence of women was forbidden.) The Greeks saw nothing wrong with the display of male genitals, with one exception. During wrestling bouts or other vigorous activities, there was a risk that the glans might become visible if the foreskin was accidentally pulled backwards. Because the sudden emergence of the glans was associated with erections and therefore with sexual activities, and because they were at pains to emphasise the non-sexual nature of their athletics, they took steps to prevent this display. They did this by tying a cord tightly around the terminal part of the foreskin, making it impossible for the glans to emerge from hiding. This cord even had a special name. It was called the *kynodesme*, which literally means the 'dog leash'. A thin leather thong, it was wound around the *akroposthion* and was either tied in a bow or tied around the waist.

This little foreskin thong can be seen on ancient Greek vase paintings, and one depiction shows an athlete in the act of tying it, in preparation for competing in some event. Some Greek authors make it clear that the wearing of the thong is primarily to enable the athlete to maintain his dignity when in the nude, and is a sign of modesty rather than an athletic protection device. Others point out that prolonged wearing of the thong will have the effect of elongating the foreskin and therefore making it even more aesthetically pleasing to Greek eyes.

Males in other cultures also dressed or decorated their penises in various ways, some gentle, others rather brutal. In New Guinea, some fiercely traditional tribal males still display their extraordinary penis gourds, an ancient custom that sees them walk about completely naked except for a long golden-yellow gourd-sheath.

In shape, the gourd, or 'horim', looks like a greatly elongated carrot, tapering slightly towards the tip. It is specially fashioned into this shape by weighing it down as it grows, so that it is stretched out longer and longer. When worn, it totally covers and obscures the penis, which is inserted into its base. The testicles, however, are allowed to hang free beneath it. It is sometimes straight and sometimes curled upwards and is held in place by a cord, the bilum string, tied around the waist. The most popular angle for the gourd is about 45 degrees above the horizontal, making it look like an erect phallus. To Western eyes it appears obscene, but to New Guinea eyes it is a proud and dignified display of adult male status, and, to make them even more conspicuous, some gourds are decorated with tassels and shells. A high-status tribesman will own a selection of gourds, some of which are only worn on special, ceremonial occasions.

The present government in New Guinea has attempted, and failed, to ban this unusual form of local costume. So far they have only succeeded in prohibiting the wearing of it in government offices.

The traditional penis decoration of the Dayaks of Borneo was a more painful affair, consisting of a hole bored left to right through the glans, into which a small polished bone could be slid on feast days. At other times a piece of wood or a feather might be inserted for everyday wear. More recently, they are said to have employed a metal rod of copper, silver or gold, with a small pebble or a ball of metal fixed at one end. This rod is about 2 mm thick and 4 cm long and, once it has been inserted, another ball is attached at the other end. This results in the glans having two little protruding spheres, one on each side. Normally these are covered by the foreskin, but, with erection, they emerge from hiding and it is said that, for the female Dayaks, penetration without this penile addition is like food without salt.

A similar form of penis decoration is mentioned in the Sanskrit treatise on love, *The Kama Sutra*, but there the penis rod is vertical instead of horizontal. This would make it more successful as an arousal device, because the upper sphere would rub against the erogenous G-spot zone on the upper vaginal wall.

In Java, tribesmen modified their penises in a slighter different manner, but one that was also calculated to appeal to their women. Making small incisions in the glans, they inserted small stones or tiny pebbles under the skin. When the cuts healed, with the stones enclosed, the tip of the penis was now covered in lumps and bumps that were said to give the women of the tribes much greater sexual satisfaction. Some modern condoms are designed to have the same effect, but without the painful surgery.

If these penis improvements seem extreme, they fade into insignificance when compared with the ritual penis mutilations found among a number of tribal peoples, especially the Australian Aboriginals. They practised a form of surgery called subincision, in which the penis of a boy when he reached manhood was first circumcised and then almost split in two.

For circumcision, the boy, aged about thirteen, was held down by an older tribseman, who silenced his cries, sometimes giving him a boomerang to bite on. Another held the foreskin and, twisting it, stretched it upwards, while yet another sliced it off with two or three cuts from a very sharp piece of broken volcanic glass. This action was watched closely by other tribesmen and it was their duty to kill the circumcisers if they failed in their task.

About four years later, the circumcised young man might then be subjected to the more serious operation of subincision. This time, the man with the knife sliced the underside of the penis down its whole length, or part of its length, slitting open the urethral tube. This wound was not allowed to close up again, so that the mutilated penis remained, as someone elegantly put it, 'kippered'. Opened up in this way, the splayed penis was much wider, and it was claimed that this gave enhanced pleasure to the females. For the males it meant that, in extreme cases, they had to urinate sitting down like a woman, because the jet of liquid now emerged downwards from the base of the penis, instead of through its tip. Some men carried little hollow tubes to improve their aim and to avoid this indignity, while others contrived to pull the scrotum up against the underside of the penis to direct the jet forwards in the normal male manner. Efficiency of sperm delivery was also impaired, although the survival of the Aboriginal tribes clearly proved that some managed to get through.

The social importance of this rite of passage was enormous. If a young man did not go through with it, he was not allowed to join his father's group of elders and was refused permission to be present at religious ceremonies. Even more important, he could not officially acquire a wife, and essentially became a social outcast. His whole tribal future was linked to the outlandish ritual of penis mutilation.

Where, one wonders, could such an unconventional idea as splitting a penis open in order to gain high social status have originated? There have been suggestions that the bisected penis is meant to look more like a vulva and the bleeding has been linked to pseudo-menstruation but this seems far-fetched.

It has also been proposed that the subincision operation is an

attempt on the part of the Aboriginals to imitate the forked penis
of the kangaroo. If this is so, the imitation is weak. The kangaroo's
penis is only forked at the tip and the subincision operation does
not create that shape. It opens up the penis into two side-by-side
halves, but it does not divide the tip into two forks.

A more likely explanation is that the operation has a mythological
origin. According to Aboriginal legends, the ritual of subincision was
given to their ancestors by a lizard-man from the dreamtime. If lizards
had a special significance to early Aborigines, they may well have
noticed that the males of these reptiles, when mating, extend what
appears to be a double penis. The copulatory organ of lizards is
composed of a pair of hemi-penes, and it may well be that the first
human penis-splitting was an attempt to imitate this condition. If the
mythical lizard-man was envisaged as possessing great powers, it may
have been felt that to copy his mating behaviour would pass such
powers on to the tribesmen.

By coincidence, the much more widespread form of penis mutilation, Egyptian circumcision, also appears to have a reptilian origin.
This is thought to have begun as part of the ancient Egyptians'
obsession with immortality. It is suggested that, when they noticed
snakes shedding their skin, they believed that the animals were being
reborn. The old body lay dry and crumpled on the ground, while
the snake emerged with a bright, shiny new body. If shedding a
piece of skin could offer the snake immortality, then it followed
that, if a human being shed a piece of skin, he, too, could become
immortal. And since the penis and the snake had a similar shape,
and there was a bit of loose skin hanging down at its tip, it was
obvious what had to be done. Remove the foreskin and, like the
serpent, you could become reborn. Patience was required, however,
because human rebirth would have to await the moment of death,
and would then take place in the afterlife.

The early Egyptians, being the elite of the ancient world, were
soon being imitated by other Middle Eastern cultures and, before
long, most of the young boys in the region were paying the painful
price of keeping up with the Egyptians. Snake worship was soon
forgotten and the only reason given now was that God preferred
circumcised penises, although why the deity should be partial to

this unusual form of child abuse was never made clear. It became so entrenched as an accepted religious practice that, despite its oddity, it managed to survive for centuries, and is still with us today. It has been estimated that about fifteen million male children are still circumcised each year, making it the most common, and most profitable, form of surgery known to man.

With advances in education in the developed world, people recently began to question the relevance of this hallowed form of sexual sculpture and the circumcisers became alarmed that their ancient ritual might fall prey to the growth of objective common sense. They defended their position by providing a more modern excuse for their delicate amputation, insisting that there were medical benefits for the circumcised male. Many scientific studies were made, searching for any possible benefits and, because of this new campaign, many doubters were encouraged to continue offering their children up for genital mutilation.

The argument for and against male circumcision has remained a heated one, with little attempt to provide a balanced view. Those against it insist that the medical arguments favouring it are all spurious and cleverly biased. They point out that modern medicine would never have dreamed up such a procedure if it were not an already entrenched religious practice. It is the only example of the surgical removal of a natural and perfectly healthy piece of tissue from the human body. If evolution put it there, why interfere with it? If an adult male wishes to have it removed for some personal reason, then let him do so. Body mutilations for aesthetic reasons are common enough. But to impose it on a healthy baby or child is inexcusable.

In the 1980s and 1990s several organisations were formed to voice opposition to circumcision. They felt that medical practice was being corrupted by greed and by religious pressures and set out to correct this.

Among their claims were that routine circumcision is equivalent to genital mutilation and has neither short-nor long-term hygienic benefits. Indeed, it has mild to severe negative physiological effects, eliciting severe pain and terror in infants and children, encoding the brain with violence. The long-term effects of the procedure include suicide and depression, with therapists arguing that men without foreskins feel a loss, relive the violence, are not 'whole', and have a diminished penis. It is also claimed that the loss of erotic tissue in the foreskin diminishes sexual pleasure.

It has been argued that circumcision increases the risk of *Staphylococcus* infection in newborn boys. *Staphylococcus* has reached epidemic proportions in many areas and has become a worldwide problem. By removing the only movable part of the penis, amputation of the foreskin, it is argued, causes increased friction during copulation, leading to micro-tears in tissue and therefore to a greater risk of HIV infection. The United States has a high rate of male circumcision and also of HIV infection. Scandinavia has a low rate of male circumcision and also of HIV infection.

The anti-circumcision lobby claims that the medical benefits of male circumcision, including reduced bladder infection, penile cancer and cervical cancer of the female partner, have been tested and found to be extremely small or non-existent. From the doctor's viewpoint, circumcision of male children violates the patient's legal rights to bodily integrity, and also violates the European Union's Convention on Human Rights. Finally, the reasons some physicians advocate circumcision of infants have been criticised as being primarily about money and to seek revenge for the pain they had when they were circumcised themselves.

The most extreme opponents of circumcision formed an organisation called UNCIRC and a book was published by the group's founder with the startling title of *The Joy of Uncircumcizing*. The basis of their approach was that, if certain tribal people could elongate their earlobes by prolonged pulling on them, it should be possible to do the same with a circumcised penis. They advocated that by various means it should be possible to restore the foreskin to something approaching its original glory. They pointed out that the penile skin is unusually elastic and that, by hanging weights on it, or taping it down in such a way that it was tugged downwards, or, if all else fails, by a surgical reconstruction procedure, the penis could be made whole again.

If men are prepared to subject themselves to such extraordinary measures, it is clear that, for some at least, circumcision is a continuing nightmare that should, in future, be stamped out completely.

The pro-circumcision lobby counter-claims that there are important medical benefits from circumcision and denies that there is any evidence of later psychological harm or loss of sexual pleasure.

Clearly these two schools of thought contradict one another on three key issues: psychological, sexual and medical. Their conclusions, based in both cases on extensive medical research, are flatly opposed to one another. Obviously, emotional factors are playing too big a part here and it will no doubt be some time before a final, unbiased, objective statement can be achieved.

In addition to subincision and circumcision, there are two other forms of genital mutilation that deserve a brief mention. One, incision, is the least damaging form of surgery and the other, skin-stripping, the most damaging.

Incision is a mild version of circumcision, in which a single cut in made in the foreskin, to expose part of the glans, but without the actual removal of any of the skin. It was observed among certain peoples of the East African coast, and on islands in Asia and Oceania. It probably developed as a token form of circumcision, permitting the coming-of-age ceremony, but with the minimum of surgical damage.

The most severe form of penis mutilation, once performed in parts of Arabia, was the brutal practice known as skin-stripping, in which the skin was removed from the entire shaft of the penis. Only young men who were able to endure this extreme form of genital torture without screaming were considered to have become respected adults.

In some African tribes, before the existence of modern condoms, a minor modification was sometimes made that enabled the men to enjoy copulation without the risk of procreation. This consisted simply of boring a small hole in the urethral tube, on the underside of the penis near the scrotum. This enabled them to ejaculate from the base of the penis instead of from the tip, so that their sperm never entered the vagina. If they wished to procreate they could place a finger over the hole at the moment of ejaculation. And they could control urination in the same way.

It is impossible to overlook a modern fashion for decorative penis modification – genital-piercing. Just as a tide of public opinion has started to sweep aside the ancient forms of penis mutilation, such as circumcision, a cross-current has appeared in the form of penis adornment.

There is nothing tribal, religious or mystical about these modern piercings. They are done purely for pleasure, either tactile during sex, or visual during foreplay. To quote a professional piercer: 'This is accomplished through added sensations provided by piercings and the jewelry worn in them. Additionally, many people find the visuals to be as exciting as the sensations. Piercings are exotic and erotic. There is a concrete physical basis for increases in sensitivity in pierced areas. Jewelry worn in piercings contacts nerve endings that would not ordinarily be accessed or stimulated.'

For those who view all decorative mutilations as abhorrent, it comes as something of a shock to discover that there are now no fewer than eight different styles of penis jewellery being attached to eager mutilees at the present time. The most traditional of these is the Prince Albert. Worn by Queen Victoria's husband, it originally consisted of a so-called dressing ring attached to the penis which was then strapped to the thigh to maintain the smooth line of the very tight trousers that were fashionable at the time. The Italian dictator Benito Mussolini also wore a Prince Albert and had a hole cut in his pocket so that he could reach it with his hand, to play with it in times of stress.

Prince Albert did not invent the piercing. It was already a well-established German custom among fighting men, used to pull their penises tightly between their legs during combat, to avoid sword injuries. It became the first type of penis decoration when the modern era of intimate piercing began in the late twentieth century. A hole is made in the underside of the penis near its tip, where there is little more than thin membrane to pierce because the urethral tube lies so close to the lower skin of the shaft. A metal ring is then inserted through the natural opening of the urethra and pushed through this hole. This can be done in both circumcised and uncircumcised males, and it takes about four weeks to heal. These modern rings, worn in looser trousers, are not usually strapped to the thigh like the nineteenth century ones.

Dolphin piercing is similar but is placed further down the shaft of the penis, 1.6 cm (0.62 inches) from the hole made for the Prince Albert. A bent bar of metal is then inserted, in through the Prince Albert hole and out through the new hole. A small, smooth metal ball is then attached to each end of the bent bar. The result is a bent bar inside the urethra, and two metal balls on the underside of the penis.

Ampallang piercing is named after a horizontal crossbar that is attached to the head of the penis of a tribal Dayak boy when he reaches manhood in Borneo. The modern version of this is in the same position, with a hole bored horizontally through the head of the penis to take a barbell, a metal tube with a small ball on either end. This creates what has been called the look of 'eyes on a beast'. This more severe piercing takes several months to heal properly.

Apadravya piercing is similar but is vertical instead of horizontal. It gains its name from its first mention in *The Kama Sutra*.

Frenum piercing goes though the surface tissue of the underside of the penis, near to its base. Some men have multiple frenums along the underside of the penis, popularly known as 'speed bumps'.

Lorum piercing is a lower frenum pushed through the linkable tissue at the base of the penis where it meets the scrotum.

Dydoe piercing goes through the rim of the glans. This type is often done in pairs and is usually only performed on men with a well-defined glans rim.

Finally, there is Pubic piercing, in which the hole is made at the point where the front of the body meets the base of the shaft, on the upper side of the penis. This is said to add extra pleasure to sex in the so-called missionary position.

One can only hope that none of these new, fashionably pierced males is ever struck by lightning.

Finally, a British male stripper called Frankie Jakeman is rumoured to have had his penis insured for $1.6 million, much to the amusement of his female partners. Although there can be little doubt that this organ plays an integral role in his theatrical performances, it is not at all clear what misfortune he feared might befall it.

Perhaps he had in mind the notorious case in 1993 of John Wayne Bobbitt, an American whose wife Lorena cut off his penis with a kitchen knife as he lay sleeping. She was angry, she said, because he would not give her an orgasm. After severing the offending appendage, she drove off and threw it from her car window. After a careful search, it was found by the police and surgically re-attached. To pay for this expensive operation, Bobbitt appeared in several pornographic films, including one called *Frankenpenis*. He later worked in a Nevada brothel and as a minister in a Las Vegas church, and his penis appears to have been in working order because he married two more times.

In Bobbitt's case it was, of course, his own penis that was surgically attached, but in China more dramatic surgery was carried out where the penis belonged to someone else. An unfortunate man, who had lost his penis in a freak accident, subjected himself to the world's first penis transplant operation. Surgeons at China's Guangzhou General Hospital performed meticulous microsurgery lasting fifteen hours to graft on a donor penis and the operation was a success. The new penis was not rejected by the man's body, but unfortunately it *was* rejected by his wife who found she could not be intimate with a stranger's sex organ. With great regret, the surgeons therefore had to cut it off and their amazing medical achievement came to nothing.

20. THE TESTICLES

Wise men have often pondered over the question of the precise location of the immortal soul. Is it in the brain? Is it in the heart? No, the answer is that it is in the testicles, because it is there that the male manufactures his sperm and it is those sperm that carry his only true hope of immortality – genetic immortality through his offspring.

Given that the testicles are such a vital part of the male body, it is fair to ask why they have been positioned so badly. By letting them dangle outside the body between the thighs, evolution appears to have been a little careless. Exposed as they are, they can easily be severely damaged by a sharp blow that strikes between the legs. Primeval hunters, grappling with an injured, struggling prey, with its bony legs thrashing about wildly, must have been incredibly vulnerable. And modern sportsmen are also all too aware of the pain that can be experienced if they are hit in this region.

Less dramatic, but also annoying, is the fact that a naked man cannot run fast without the discomfort of his testicles flapping from side to side. It was this apparent design fault in the male body that probably led to the invention of the first ever human garment, the primitive equivalent of the jockstrap. Once held tight in this way, however, a new problem arose – the chafing of the loose scrotal skin, leading to sores and irritations.

There is no getting away from it: internal testicles of the kind found in many other animals would have been safer and more comfortable for the newly bipedal, evolving human male. So the obvious question is: why do men have external testicles?

The traditional answer, and one that has been around for many

212

years, is that having them outside the abdomen keeps them slightly cooler, and this slight drop in temperature favours sperm production. Many research projects have been carried out in support of this theory.

One authority states that the ideal temperature for sperm production is about 3 degrees Celsius lower than normal body temperature. Another claims that ten minutes in the sauna for ten days has been shown to decrease sperm production by a dramatic 50 per cent ten weeks later.

Other activities or habits said to increase the temperature of the testicles in a detrimental way include working with a laptop computer on your lap, driving a truck on long journeys, sleeping on a heated waterbed, intense exercising, or wearing very tight underwear or trousers.

These are not idle suggestions; they are the result of careful investigations. The laptop example, for instance, is based on a study of twenty-nine young men who were asked to sit for an hour working at their laptops while balancing them on their thighs. This produced a rise in scrotal temperature of as much as 2.8 degrees Celsius. In other words, it made the testicles almost as hot as if they were internally positioned. Only part of this was due to the heat coming from the laptop itself. When the men were asked to sit with their legs held tightly together, but without the laptops on top of them, there was still a significant rise in temperature, as much as 2.1 Celsius. So the act of balancing the laptop had an even greater effect than the heat actually generated by it.

In other words, the primary cause of testicle heat rise is squeezing the testicles between your legs, either by posture or by tight clothing. Any activity, whether at work or play, that involves prolonged sitting with the legs pushed together, is likely to reduce sperm production. If the testicles are allowed freer movement, then a muscle that controls how high or low they hang will adjust to the existing temperature. When the heat rises, the muscle allows the scrotum to hang lower and permits ventilation. When it gets colder, the muscle contracts and the testicles rise up into a snug, warmer position. In this way, the unrestricted testicles are kept at the exact temperature that makes sperm production fully efficient.

The muscle that retracts the testicles when a man is cold also does so when he is under stress. This is because, in a primitive situation, a sudden increase in stress was associated with possible physical danger, when the testicles had to be protected from harm.

Curiously, the other time when the muscle contracts and raises the testicles close to the male's body is just before he ejaculates. The only explanation for this is that, if both partners are reaching a sexual climax together, there may be enough convulsive thrashing of limbs to endanger the testicles, were they hanging more loosely.

One of the problems the testicles face is fitting in between the male legs. There is not much room there, especially if the male in question has athletically robust thighs. Evolution has solved this problem by making one testicle hang lower than the other. In this way, they need less room. If they were to hang down to exactly the same degree their symmetry would require more space.

In the majority of men, the left testicle hangs lower than the right. One study, involving 386 men, puts the figure at 65 per cent left-low, 22 per cent right-low, and 13 per cent equal. At the Kinsey Institute for Sex Research, a much bigger study was made, using no fewer than 6,544 men, and this time the figures were more dramatic: 90 per cent left-low, only 5 per cent right-low and 5 per cent equal.

Nobody seems to know why only 10 per cent of men lack the lower left testicle feature, any more than we know why only 10 per cent of men are left-handed, but there have been suggestions that these two groups are somehow related. The relationship is a tenuous one, however. The simple idea that the 90 per cent of men who are right-handed also have a lower left testicle does not work. The more subtle truth is that the likelihood of having a lower left testicle is very slightly greater in right-handed men than in left-handed men. This is such a weak connection that, for the moment, the lopsided bias of the testicles must remain a mystery.

For body-building muscle men with gigantic thighs, the lack of space for their testicles sometimes becomes acute, but help is at hand. The anabolic steroids that they usually take to increase dramatically the size of their muscles also have the effect of reducing their testicles to about half normal size. So, as the body-builder's muscles begin to swell, his testicles start to shrink, thus making it easy to accommodate them. The reason this happens is that the steroids mimic the effects of testosterone, and, since the testicles are the main source of this male hormone, they react as if there is no need to continue production. When the body-builders stop taking the steroids, their testicles soon return to their normal size.

When they are working efficiently, the 2-inch-long, egg-shaped human testes produce about two hundred million sperm every twenty-four hours. The testes contain many narrow ducts and it is inside these that the sperm are actually created. Around these ducts are cells that produce the male sex hormone called testosterone. The sperm are minute; five hundred laid head to tail would only stretch a distance of one inch. Once they have been produced they are stored in readiness for possible ejaculation. When the male's sexual climax is reached, they are mixed with fluid from several glands and jettisoned down the urethral tube inside the penis.

If, during foreplay, the male sexual climax is deliberately delayed for a long period of time after full erection has been achieved, the preparatory stages of sperm readiness are unnaturally prolonged to a point where the testicles begin to ache. This condition is familiar enough for it to have acquired the slang name of 'blue balls'.

If a male is not sexually active, the build-up of sperm may result in nocturnal emissions, usually called 'wet dreams', in which he ejaculates spontaneously in his sleep and releases the tension in his sperm store. If there are no ejaculations, the unused sperm will eventually be reabsorbed. This does not help his reproductive system to run smoothly however, and it has been shown that regular ejaculation improves sperm production. It therefore follows that, for a male who is temporarily inactive, masturbation would help to keep his system in good shape.

There is a wide variation in the number of sperm released with each ejaculation – anything from two million to two hundred million. A figure of sixty million is usually given as the minimum required for successful fertilisation, even though only one of these will actually enter the female egg and fertilise it. The large number is needed to create the perfect chemical field in which the ultimately successful sperm can travel on its long journey.

Because so many studies have proved beyond doubt that the efficiency of sperm production is increased when the temperature of the testicles is slightly lowered, it has been taken for granted that this provides a full answer to the question: why are the testicles placed in such a vulnerable position outside the male body? Clearly, they are put at risk because it makes them more efficient. Every textbook has repeated this concept until it has become accepted as common knowledge. Unfortunately it does not fit the evolutionary facts. As a full explanation, it has two major flaws.

The first is that, in very hot climates, such as tropical Africa where our ancient ancestors evolved, the air temperature is often so high that having external testicles would not help. Indeed, in some of the most intensely hot regions, where the thermometer can climb as high as 58 degrees Celsius (136 degrees Fahrenheit) the external testicles may be hotter than if they were snugly internal. Secondly, when other mammalian species are examined, it turns out that, although many do have a scrotum containing external testicles, many others do not.

If so many species can breed successfully in a hot climate, or with internal testicles, there must be some other advantage in risking them in the outside world. So far, only one suggestion has been advanced to explain this, the Michael Chance Concussive Theory. The idea came to the eponymous another when he read the report of urine tests carried out on the Oxford and Cambridge boat race crews. As with many sporting events, these tests are carried out to check the use of forbidden drugs, but what Michael Chance spotted was something altogether different. The report stated that the urine taken after the race contained prostatic fluid that was not there before the start of the race.

In other words, one of the internal glands that supplies the seminal fluid in which the sperm are ejaculated was leaking under the intense muscular pressure of the rowing action during the famously demanding boat race. Each stroke of the oar would cause sudden, concussive pressure inside the oarsman's abdomen. Unlike the bladder, the male reproductive tract has no sphincter, and the repeated pressure on the prostate gland by the vigorous rowing actions had resulted in its contents being squeezed into the urethral

216

tube, where it was picked up by the urine expelled for the post-race tests.

This made Chance realise that, if human males had internal testicles, these too would be squeezed by any powerful, sudden pressure, and this could result in the serious loss of sperm. The hunting activities of tribal males would inevitably involve such pressures, as the hunters leapt, jumped and generally did battle with their large prey. If their testicles had been internal, the repeated tensing and squeezing of the abdominal muscles would have led to a loss of sperm as well as to a loss of the seminal fluid. But with external testicles, the sperm would be immune from such pressure changes and would not be needlessly wasted.

On this basis, it should follow that any mammalian species, regardless of the temperature of the environment in which it lived, should have internal testicles if it led a quiet life, and external ones if it led a life full of sudden, concussive pressures. In other words, a crawling, walking or burrowing species should have internal testicles and a leaping, racing, thumping one should have external testicles. And that is exactly what he found when he studied a wide variety of species, confirming his idea and relegating the cooling principle to the level of a secondary adaptation.

To give just two examples, all male anteaters and armadillos enjoy a non-concussive lifestyle and they all have internal testicles, even those that live in the very hot tropics. All male horned animals – cattle, goats, antelope, deer and the rest – must endure considerable concussive pressure, especially in the rutting season, when they bang their heads together in furious contests of strength, and they all have external testicles, again regardless of where they live.

One of the unfortunate consequences of this evolutionary development is that it is easy to castrate the human male. The removal of internal testicles would require specialised surgery. Performed crudely, it would cause death. But with external testicles it is simply a matter of a quick snip. As a result we have, in the past, witnessed the appearance of two special categories of human males, the castrati and the eunuchs.

The castrati were victims of the Roman Catholic Church's refusal to have women singing in their choirs in the sixteenth century. The priests loved the pure, high tones of the small boys in the choir and wanted to keep them singing that way for as long as possible. Sadly, when their voices broke at puberty, the beautiful ringing tones were lost. The solution was to select certain small boys before they had reached puberty and cut off their testicles. By removing the source of testosterone in this way, the development of adult masculine features was prevented and as the boys grew older their voices retained the high, pure quality that the priests so enjoyed. Furthermore, with growing physical stature, the adult castrati developed more powerful voices, giving them a unique sound unlike that of small boys, adult males or adult females. It was this sound that thrilled the priests and kept them snipping away at testicles for more than three centuries.

Tragically for the little boys, only about 1 per cent of castrati ever developed into successful adult singers. But for those who did there was a life of luxury and adoration. They were the pop stars of their day, branching out successfully from church choirs to operas, and treated as great celebrities wherever they went, with large entourages to tend to their every need, and crowds of admiring fans.

It was not until Victorian times that the Roman Catholic Church finally outlawed this gross form of sexual abuse of children. The last mutilation was performed in Italy in 1870, and in 1902 the Pope banned the practice forever. Some of the older castrati were still singing at this time and the last one to be heard in a church choir finally left in 1913.

The making of eunuchs for other reasons has a much longer history. The earliest recorded instances date from more than four thousand years ago in ancient Sumer, and the practice is still occurring today in certain parts of the world. The main reason has always been to create adult males who cannot enjoy procreation and therefore pose no threat to dominant, rival males.

If a tyrant became so powerful that he could assemble a harem of breeding females, he had to face the problem of protecting them and preventing other sexually active males from gaining access to them. This required strong male guards, but how could these be trusted? The answer, of course, was that they could not be trusted and the only solution was to render them impotent.

To this end, young boys were castrated and then pressed into service. The compensation for their genital loss was that they were so valuable to their masters that they had a secure job, and could enjoy an unusually comfortable lifestyle. Their primary task was to ensure that the harem females were faithful to their master and this is how they came by their special title, eunuch meaning literally 'bed-keeper'.

According to some legends, young men who were not castrated practised hard at retracting their testicles into their abdomens. Once they had achieved this unusual feat they were able to pass themselves off as eunuchs and were then allowed to enter the communal bath-house on days reserved for protected women of high rank. What happened next is not recorded but can easily be imagined.

In addition to their role as harem guards, eunuchs were also widely employed as servants of powerful rulers and ancient courts. They were more trusted than ordinary male servants because they had no family loyalties or ambitions. In the Ming Dynasty in ancient China, there were no fewer than 70,000 eunuchs acting as servants in the Imperial Palace. Their role was considered so desirable and the rewards so great that self-castration became common in order to gain access. This became so widespread that it was eventually made illegal. As the centuries passed, the number of eunuchs in the Palace dwindled and when the practice was finally stopped in 1912, by coincidence at about the same time as the demise of the castrati in Italy, there were fewer than five hundred of them left.

When the Chinese practice of castrating was halted, the surviving eunuchs lived on as best they could in a world that no longer valued them. By the 1960s there were fewer than thirty, and the last one died in 1996 at the advanced age of ninety-three.

Today, the only place where large numbers of eunuchs can be seen is India, where there are currently about a million of them. They are known as *hijras* and have several sources of income. One is aggressive begging, in which they roam the city streets accosting people and threatening to show them their operation scars. To avoid embarrassment people pay them to go away.

Hijras have also carved a social niche for themselves by attending special social occasions such as childbirth and weddings, and insisting on blessing the proceedings, whether people want their blessings or not. The eunuchs, who live together in special ghettos, have spies everywhere to warn them of an impending birth or marriage and, once the great day has arrived, they converge on the place, sing and dance, and then demand large sums of money as a fee for giving their blessings. If they are not paid, their blessings become curses, and the more superstitious of their victims fear this and are usually persuaded to part with hard-earned cash to protect their new infants or their marriages.

The making of an Indian *hijra* is more extreme than with most other eunuchs because he loses his penis as well as his testicles. The boy is doped with opium and a cord is tied tightly around the genitals. Then the penis and testicles are sliced off with a single cut of a sharp knife. After the wound has healed, the boy is taken under the wing of a guru, who will look after him and care for him until he is fully adult. For the rest of his life he will be given a female name and referred to as 'she'. And he/she will probably earn extra money as a prostitute serving homosexuals, or perhaps heterosexual men who cannot afford a female prostitute.

These modern eunuchs refer to themselves as the third gender of India, but it is hard for foreigners to understand why any boy or young man would want to subject himself to a major mutilation to join such a group. The answer seems to lie in their strength-in-numbers strategy. Most of them begin life with homosexual tendencies and find themselves damned by society. Unable to face the prospect of marrying and setting up a family, they are left with two alternatives: either remain an isolated outsider, scorned by society, or join the ranks of a huge, parallel society of eunuchs. The act of castration then effectively becomes an initiation ritual, consigning them, permanently and irreversibly, into a lifelong membership of this special, distinctive group.

In recent years, with more liberal attitudes, there have been attempts to raise the lowly status of the eunuch community. There have even been *hijra* beauty contests and fashion shows and their future looks brighter. If this trend continues, the time may one day arrive when the gay community of India no longer feels the need to cut off their testicles in order to belong to their special community.

On another topic, there is a popular belief that the word 'testament' is derived from 'testicle' – an idea that has upset some pious biblical scholars. The truth is that both words were derived from the Latin *testis*, meaning a witness. This root word *testis* has given rise to a whole collection of modern words, including testify, contest, testimony and attest, in addition to testament and testicle. All of them have a relationship to the idea of bearing witness.

It is known that, in pre-biblical times, men sometimes swore an oath and gave witness while placing a hand on another man's testicles, the idea being that the testicles bore witness to a man's virility. In Genesis there is a coy reference to this when 'the servant put his hand under the thigh of Abraham his master and swore to him concerning that matter'. Later, in Roman times, men were reputed to place their right hands on their own testicles and swear by them before giving testimony in court. The reason for this is that, in ancient Rome, eunuchs and women were not allowed to testify in court, and men had to provide proof that they were 'entire'. Roman law demanded that no man could bear witness unless he possessed both testicles. A prosecutor could demand evidence of this and there was an old legal saying that went: *Testis Unus, Testis Nullus*, meaning if you have only one testicle your testimony is null and void. Today this phrase is still used in law, but the meaning has been changed. It is now taken to mean: 'A single witness is no witness'.

A brief mention must be made of a popular legend concerning the Pope's testicles. For nearly two thousand years it has been a strict rule that no eunuch or woman could be elected to the elevated role of the Holy Father. The pontiff must possess two testicles and this fact must be established beyond doubt. It is not clear why a celibate pope should need two testicles, but it presumably has something to do with the sexist bias of the Roman Catholic Church. In order to prove the existence of a pair of papal appendages, part of the election process of a new pope sees him sitting on a special 'testicle chair' with a hole in the middle of its seat. This seat was either an old Roman birthing chair or one specially designed to allow the great man to, as it were, lower his undercarriage.

Once he is seated in this chair, his testicles are formally examined. According to one version of the story, the cardinals file past, each taking a peek to satisfy themselves that he is complete; according to another, a designated cardinal thrusts his hand under the chair and feels the testicles. When he confirms that there are indeed two, he shouts out in a loud voice. *'Testiculos habet'* ('He has testicles') and all the clerics reply, *'Deo gratias'* ('Thanks be to God'). Some authors say that the exact phrase shouted out by the cardinal is *'Testiculos habet et bene pendentes'* meaning, 'He has testicles, and they hang nicely'. It is claimed that this ritual has been taking place for more than a thousand years and that it was originally introduced because, according to the Law of Moses, eunuchs could not enter into the sanctuary.

The story has been embellished, probably as an anti-papal satire, with the legend of Pope Joan, a young Englishwoman who is said to have travelled to Rome in the ninth century where she disguised herself as a man and managed to get herself elected Pope. Her deception was discovered when she rather inconveniently gave birth during a papal procession. The angry crowd then stoned both her and her newborn to death. After that scandal, Pope Joan's brief rule of two years, four months and eight days, from the years 855 to 858, as Pope John VIII, was expunged from the record, and it was decided that, in future, the cardinals would take no chances: the testicle chair was duly introduced. Today, this tale is recognised as pure fiction, but the basic facts about the ritual of intimate papal examinations seem to be generally accepted.

Finally, there are two scrotum-piercings that have become popular in recent years, as part of the increasingly fashionable obsession with decorative body mutilation. First, there is Hafada piercing in which a ring is attached to the scrotal sac. It is usually placed to one side, although it may also be inserted on the midline. Sometimes a pair of rings is added. The name is said to derive from an Arab rite of passage that takes place at puberty.

Second, there is Guiche piercing, named after a puberty ritual said to take place in certain parts of the South Pacific. This is a horizontal piercing that is positioned at the back of the scrotum at the point where it meets the perineum. Although this type of metal adornment is said by those who have submitted themselves to it to be pleasurably erotic, it would not seem to be advisable for men who, for example, are participating in the Tour de France.

21. THE BUTTOCKS

Human beings are the only animals to display a pair of muscular, hemispherical buttocks. If we were looking for a name to distinguish us from all the other primates, an alternative to the Naked Ape would be the Round-bottomed Ape, thanks to the pair of powerful *gluteus maximus* muscles that we all possess on our rumps.

Our buttocks evolved in response to our newly developed vertical posture, enabling us to remain upright for long periods of time. They should therefore be viewed as a noble part of our unique character but instead more often than not they are the subject of ridicule. Perhaps if the anus were situated elsewhere on the human body we would treat them with a little more respect, but as it is they are the butt of endless jokes.

However, when a group of women were asked which part of the male body they found most exciting, the majority of them had to admit that it was this region, rivalled only by the eyes. A taut, powerful bottom, casually displayed by an athletic man as he walks past a woman, is the focus of her gaze, although she might be reluctant to admit it.

One young woman who had the courage to risk ridicule and sing the praises of the male buttocks was the Californian actress Christie Jenkins. In 1980 her book *A Woman Looks at Men's Bums* was published, complete with an array of male posteriors photographed by the author herself. To a man, these pictures seem unremarkable, just a collection of male rear ends, but to a woman, apparently, they have enormous appeal if only she will admit it.

The author claims she became such a connoisseur that she could tell the athlete from the occasional jogger from the desk-bound executive at a single glance, and therein lies the secret. It is the taut, muscular buttock that has sex appeal for women, and the more powerful it is the higher she rates it. There are two primeval reasons for this. Firstly, the muscular buttock symbolises the healthy fitness of a successful hunter, one who can run and leap and chase, and bring home the kill to feed his family. And secondly, the muscular buttock suggests a man who will be capable of impressive pelvic thrusts, thrusts powerful enough to satisfy a woman's sexual needs. Soft, flabby buttocks may belong to an intellectual genius, but he will always lack this primitive appeal.

As with other parts of the male body, help is at hand today to improve the shape and firmness of the buttocks. Cosmetic surgeons offer a Buttock-Modelling service, or Buttock Augmentation, that involves inserting special implants to replace the missing natural volume. They insist that these implants cannot be felt and do not interfere with movement, and that they provide a completely natural feeling when sitting. No scars are visible because the implants are inserted through incisions made in the vertical cleft between the buttocks.

These buttock implants are made from solid silicone, more resistant, it is said, to wear and tear, or a softer form of cohesive silicone encased in a silicone bag that feels more natural but may rupture if put under great pressure.

For men who are reluctant to resort to surgery to improve their behinds, there is always the less drastic measure of dressing appropriately. This has a long history, dating back to the fourteenth century. According to fashion historians: 'At this time noblemen abandoned their long robes and, for public appearances, wore short doublets or padded jackets that fitted close to the body with leg-hugging hose revealing and emphasizing the thighs and buttocks.'

In modern times, the wearing of tight trousers or jeans without the covering of a jacket has had a similar effect and the fashion world has even at one point introduced padding in the form of bottom falsies to swell out the buttock region, although this seems to have been favoured more by women than by men.

Because buttocks are uniquely human, it was assumed in earlier times that the Devil lacked buttocks and images of him sometimes revealed that, in place of buttocks, he had another face where his rump should have been. Deeply superstitious people believed that if evil forces, controlled by the Devil, were threatening them, the most effective form of defence was to show the Devil one's wonderfully rounded behind, thus tormenting him by reminding him of his weakness. The idea was that he would be so envious that he could not bear to look upon this uniquely human part of the anatomy and would therefore have to avert his gaze. This would then protect the superstitious 'mooners' from the terrible threat of his Evil Eye.

These early examples of mooning carried no joke message. They were deadly serious. Even Martin Luther is said to have employed this method of protection when he was tormented by visions of the Devil, and some early Christian churches were decorated with stone carvings showing figures displaying their naked buttocks to protect the buildings from the Evil Eye. If there was a particularly terrifying storm at night, with lightning and thunder caused, it was believed, by the Devil's anger, some men would rush to their front doors and thrust their naked buttocks in the direction of the storm in an attempt to drive it away.

Modern mooners are a different matter altogether, and it is doubtful whether they have any knowledge of the more serious activities of their predecessors. For them, the act of lowering their trousers and thrusting their naked buttocks at startled or amused onlookers is no more than a daring taunt or insult, a partial form of streaking. It is not a sexual gesture because the genitals are always carefully hidden from view. It is more related to the idea of defecating on one's victim, or threatening to do so.

This modern version of buttock display began in 1968 when American college students began to moon from windows to shock passers-by. Since then the custom has spread and has developed its own terminology. For example, mooning with one's buttocks pressed against a car window is called 'pressed ham' while mooning on a very cold day is called 'blue mooning'.

The legality of mooning has often been questioned. In countries where exposing the genitals is illegal there have been attempts to outlaw it, but, since the genitals are never exposed, this has created something of a headache for the legal profession. Recently, a court in Maryland reached a decision that 'mooning is a form of expression protected by the United States constitutional right of freedom of speech', though it is not clear precisely what words the buttocks in question were uttering.

Although American students initiated the modern craze for joke mooning, it is also known from other regions and other cultures. In New Zealand, the Maori moon as a sign of disrespect. To them the buttocks are a taboo area and showing them to a victim is therefore insulting. When Queen Elizabeth II visited New Zealand on one occasion, a Maori mooned her as a protest and was promptly arrested and charged with indecent exposure. He defended himself by saying that he was performing a traditional sign of protest and that it was a part of his culture.

The Queen received similar treatment back home in London in 2000 when a group of anti-monarchists organised a 'Moon Against the Monarchy' event outside Buckingham Palace. Large numbers of police were drafted in to protect royal eyes from this unusual display of massed buttocks, and this deterred many of the would-be mooners, but some persisted and were arrested.

In the sporting world, the act of mooning can be expensive. When an American footballer mooned at rival supporters after scoring a touchdown in 2005, his defence was that he was only retaliating against fans who used the same gesture towards his and other opposing teams when their buses were leaving the stadium, but he still had to pay a fine of $10,000.

For Marlon Brando an improvised act of mooning brought him, not an arrest or a fine, but a round of applause from a film crew. Towards the end of the notorious film *Last Tango in Paris*, Brando is seen leaving a stuffy dancehall where elderly couples are studiously dancing the tango. The director was unhappy with the exit and asked Brando to do something outlandish. In the next take, without warning, Brando dropped his trousers and pushed his bare buttocks into the face of the matronly lady who was in charge of the dancehall.

Because of the thickness of the *gluteus maximus* muscles, the buttocks have often been chosen as the most suitable site for administering corporal punishment. The flesh of the buttocks can be painfully assaulted without any serious damage to the underlying bones, or other organs. Bottom-spanking was the normal form of punishment in European schools in earlier times and more serious buttock-flogging has been a widespread legal punishment in the past. In some countries it still occurs today, and there was a recent case in Singapore where it was announced that an act of vandalism by an American teenager would be punished by six strokes of a rattan cane.

Back home in the United States, voices were raised in anger about this cruel form of punishment and President Clinton intervened, requesting a more lenient sentence for the young man. Singapore's Head of State responded by reducing the sentence to four strokes of the cane, as a gesture of goodwill towards the American President. This was not quite what Clinton had expected, but he had to accept it.

In Singapore, the teenager was stripped naked in the prison caning room. His arms and legs were fastened to a trestle by straps, and the caner, using his full body weight, struck the boy's buttocks with a 13-mm-thick rattan rod, which had been soaked overnight to prevent it from splitting. Each stroke came about half a minute apart. After the final stroke, the teenage boy shook hands with his caner and walked back to his cell unaided.

The next chapter in this story of beaten buttocks was surprising. No fewer than seven different states, across America, proposed laws that imitated the one in Singapore. In each case, it was proposed that young vandals should be beaten for their crimes. In California there was an attempt to have a law passed that would punish graffiti artists by having them beaten on the buttocks with a wooden paddle. In Tennessee a move was made to introduce public buttock-caning on the courthouse steps as a way of punishing local vandals. None of these proposals ever became law, although in some instances the voting for and against was very close.

In addition to Singapore, fifteen other countries still insist on caning their criminals. These include Malaysia, Pakistan and Brunei. The argument against caning, in the rest of the world, is essentially that violence creates violence. Those who are beaten will one day seek their revenge and, far from preventing crime, violent punishments will increase it. A comparison between the high crime rate in the United States and the low crime rate in Singapore would seem to refute this, but the matter is not that simple. Other countries that have banned corporal punishment have seen no rise in crime following this step.

There is another complication. Beating may lead to a sullen, deep-seated resentment that ultimately sees violent reprisals taken against society, but it can also have precisely the opposite effect, creating a masochistic pleasure in being beaten. Painful rituals involving the spanking, caning or whipping of the buttocks are common features of the bizarre world of the sado-masochist. Specialist clubs exist in most major cities where masochists can gain sexual pleasure in this way, and such clubs seem to have been most popular during the period when the caning of schoolboys was a widespread practice. This underlines the sexual nature of all forms of buttock punishment and may explain why, even in Singapore and Malaysia, female criminals are excused this form of punishment.

One self-confessed spanking fetishist confessed that, when he was being beaten during adolescence, he experienced his first sexual arousal. He felt that the flesh of his buttocks was somehow connected to his brain along the same neural pathways that served his genitalia. As he experienced the pain of each blow to his buttocks he simultaneously experienced sexual pleasure.

Psychoanalysts have seen an unconscious connection between the rhythmic blows of a spanking and the rhythmic thrusts of a penis during copulation. To explain why spanking is so much more violent than other forms of sexual contact, it has been suggested that 'the sexual arousal nerves in the bottom are buried in a layer of fat and require harder stimulation to trigger them'. Whatever the truth of the matter, it has to be admitted that sex and pain in the buttock region can never be totally separated and, although the psycho-analytical description of spanking as symbolic rape is a trifle over-stated, it cannot be completely ignored.

Finally, there is a recent report, hard to believe, that doctors at a Polish hospital have exploited a man's buttocks in an entirely new way. The young man, who had been suffering from cancer of the tongue, had it surgically removed and replaced with a new tongue modelled from skin, fat and nerve tissue collected from his buttocks. His surgeon reported: 'The new tongue is alive and well-supplied with blood, and the patient is doing well.' What its first words were has not been recorded, but headline writers inevitably had a field day with such phrases as 'talking through his backside' and 'tongue in cheek'.

22. THE LEGS

Human legs are unique. It is important to remember that, when we talk of our legs, we are in reality talking about our hind legs. We tend to forget this because, like birds, we have stopped walking on our front legs and have used them for other things. Among all the four thousand different species of mammals alive today we humans are the only ones to walk and run on their hind legs throughout their entire adult lives. Some other species can do this briefly, from time to time, but for them it is never a primary method of locomotion. Gibbons, for example, can walk bipedally in a clumsy way when they are on the ground, but they spend almost their entire lives up in the treetops. Other mammals, such as chimpanzees and bears, may occasionally rear up and take a few faltering steps on their hind legs before reverting to their normal four-legged gait, and pet dogs can be taught to do the same, but only human beings can truly be called bipedal walkers. Kangaroos and wallabies are truly bipedal, of course, but they are hoppers not walkers. Only man stands erect and then strides out, or runs, as a primary form of locomotion.

This unique way of life put a huge demand on our lower limbs. Before, when our remote ancestors were four-footed, the hind legs only had to bear half the burden of moving about. Now they had to take it all. As a result, during the course of evolution, they had to become stronger and longer. The long straight human leg accounts for half the body's height. When artists are sketching the human form they divide it up into four roughly equal parts: from the sole to the bottom of the knee-cap, from the knee-cap to the pubic

231

region, from the pubic region to the nipples and from the nipples to the top of the head. This is the adult shape. The proportions of children are slightly different, with the legs being shorter in relation to the upper body.

The foundation of our powerful legs comprises four bones: the massive thigh bone, the longest bone in the human body, called the femur; the knee-cap which protects the front of the hinge joint at the base of the femur, called the patella; the shin bone, which articulates with the femur, called the tibia; and the splint bone, which lies alongside the tibia, called the fibula.

Propelled by his well-muscled legs, the human male has sailed nearly 8 feet up into the air and has managed a long jump of nearly 30 feet. Top sprinters can reach speeds of 27 miles per hour. And long-distance runners can cover twenty-six miles in about two hours. Marathon dancing has dragged on, week after week, with the participants in a state of near exhaustion, for as long as 214 days. Such feats of strength and endurance are a remarkable testimony to the evolution of the human legs during a million years of chasing and hunting. No wonder the legs have come to be thought of as symbols of stability, power and nobility.

Some sportsmen have tried to improve their leg performance by sacrificing a little of their masculinity, namely their leg hair. It is believed that hairy male legs are slower than smooth ones in certain contexts. This seems improbable, but male leg-shaving is a small price to pay for a possible time advantage, especially where races can be won or lost by a fraction of a second.

The two sports where leg-shaving is common are cycling and swimming. Cyclists do this, they say, because road-rash heals faster without the presence of leg hair. Some admit that they do it for purely aesthetic reasons. Others say that it feels faster when they cannot sense the wind resistance in the hair on their naked, straining legs. Swimmers shave their legs, and other parts of their bodies, too, because it is supposed to reduce friction in the water.

Away from the sporting arena, legs are also conspicuous in erotic contexts. Because adults possess legs that are both relatively and in absolute terms longer than those of children, it is inevitable that long legs should be equated with sexuality. A long-legged man is considered more sexually appealing than a short-legged one. Those with unduly short legs are often given insulting nicknames such as 'short-arse', and are sometimes driven to take extreme measures to appear taller than they really are. Male movie stars with short legs often wear lifts, tall shoes that raise them up a little. And in extreme cases they have stood on hidden boxes or required their female co-stars to walk beside them in a hidden trench.

In addition to simple leg length, male leg appeal also depends on the ratio of torso to legs. When photographs of men with varying leg-to-torso ratios were shown to a large number of people, and they were asked to rate them for sex appeal, it was discovered that the ideal male ratio was 1:1. The man with legs and torso of the same length was considered to have a more powerful build and therefore to be more attractive. The perfect man, therefore, is the one who has long legs and a long torso to match them. A shorter-legged man, who has a short torso to match, will also be seen as attractive, but only if he is viewed by himself. As soon as he stands next to a taller woman, his appeal is reduced.

In tribal societies in the tropics, men's legs have usually been exposed simply as a response to the heat, but a new situation arose when human populations grew and began to become more complex. From ancient civilisations to the present day, most males have worn some sort of leg covering. Ignoring all the subtle whims of fashion, these coverings can be crudely classified as:

Long and undivided leg covering that was worn in ancient civilisations by the majority of men, and is still worn today by sheiks and popes, by Arabs and monks and celibate clergy.

Long and divided leg covering that was worn in ancient times by some nomads and horsemen. In the sixteenth century it was worn as hose, later becoming tight breeches. Then, after the French Revolution, it was replaced by the loose trousers of peasants and is worn today by almost all adult males, as trousers or jeans.

Short and undivided leg covering that was worn in earlier times as the Greek chlamys, and is still worn today as kilts by Scotsmen and as short skirts by festive Albanians.

Short and divided leg covering that is worn traditionally by Alpine people as lederhosen, and by men from many cultures as shorts, when taking part in athletics or sports, or on holiday.

As far as erotic displays were concerned, there were only two themes in male leg clothing, the very tight and the very short. The tight hose, or breeches, revealed the shapeliness of the leg, and the very short garments exposed the naked lower legs. All kinds of long and loose clothing concealed the male legs and were therefore preferred in puritanical ages.

Tight hose began in the fourteenth century as a pair of long stockings, a fashion that outraged the Church. Later, in the fifteenth century, the stockings were joined at the back. Later still, they were also joined at the front, creating what we would today call tights. While this was happening the tunic, a doublet, was getting shorter and shorter, until there was a risk of exposing the genitals. It was at this point that the codpiece was introduced, first as a separate item of clothing and then, later still, as part of the hose. This new type of hose was usually referred to as breeches.

European aristocrats had become addicted to these breeches when the French Revolution arrived and swept away the older fashions. Now a new peasant look was required and this is the point at which the loose working trousers of peasants entered the scene. Their popularity has survived right up to the present day, when the vast majority of men are still striding out in comfortable trousers or jeans. They may not show off a shapely male leg, or expose an erotically hairy knee, but they do appear to suit most men, who no longer seem to care about showing off their fine limbs to their female partners. Only modern sportsmen and athletes in their flimsy shorts are today routinely prepared to expose their leg flesh to female eyes. Bearing in mind that most other males, with their cars, desks and computers, are now physically more inactive than they have ever been before, this is probably just as well.

In the twenty-first century, attempts by gay designers to introduce daring new styles in male leg fashions have rarely met with any success. They may have looked amusing on the catwalk, but they have never made it to the high street. Crumpled trousers and grubby jeans still reign supreme in the world of the manly male. Ignoring male preferences, the designers soldier on. As one critic put it: 'More designers are designing, without kindness, exclusively for the thin teenager: skinny, pretty, elegant fashions for boys with legs like cooked spaghetti. Effeminate fashion bound to alienate a lot of blokes older than 30.'

Short trousers have a very curious history. They began as part of the uniform of the Boy Scout movement in the early twentieth century. From this beginning they spread out in two directions, into sport and into boys' schools. In schools they created a whole new coming-of-age ritual, when boys, who had been wearing short trousers all through their childhood, were finally given their first pair of long trousers. This concealed their soon-to-be hairy legs from view and transferred them into the category of sexually mature adult males. Today this sharp distinction between the short trousers of youth and the long trousers of adulthood has been blurred by increasingly varied fashion styles.

In the sports arena nowadays almost any amount of male leg exposure is permitted, although we have never seen a return to the full nudity of the ancient Olympics. In the nineteenth century the rules were much stricter. Football players took to the field wearing thick woollen stockings and three-quarter length trousers that reached down over the tops of the stockings, or were tucked into them. By the end of that century, the trousers ended just above the knee and the stockings just below it, allowing the daring exposure of a pair of knobbly knees. During the twentieth century the trousers became progressively shorter, and at one point in the 1970s they became so short that they exposed most of the footballers' well-developed thighs.

Similar changes took place in other sports such as tennis, but American football took a different route, displaying the shapely legs of its players by returning to the older strategy of tight breeches. Only the slower games, like cricket, retained the old-fashioned style of loose leg cover-up.

Back in Victorian times, the exposure of any naked male leg, man or boy, was considered too provocatively sexual to be permitted. So intense and complete was this suppression of the erotic leg that even the word itself became prohibited in polite circles. In the United States legs were called limbs. Other euphemisms included 'extremities', 'benders', 'underpinners' and 'understandings'. At table a chicken leg became dark meat.

One bride was so appalled by the sight of her husband's exposed legs when he appeared in his nightshirt on their wedding night that she spent the rest of her honeymoon making him long nightgowns that would cover the offending flesh completely. It is hard for us to comprehend a social climate in which such extremes of prudery could flourish, but the fact remains that legs were a taboo subject for a very long time, confirming their erotic potential.

Finally, those individuals whose whole career is focused on their legs often live in fear of an accident. A broken leg is horrible for anyone, but for them it could spell the sudden end of a lavish lifestyle. As a precaution some of them have taken out insurance policies to protect themselves from such a disaster. Fred Astaire had each leg insured for $75,000 which, today, seems very little. It fades into insignificance alongside the £25 million that the *Lord of the Dance* star Michael Flatley would have been paid had someone crippled him. Top footballers' legs are also immensely valuable: the Brazilian Ronaldo had his insured for $26 million.

23. THE FEET

It has been said that man stands alone because he alone stands. To put it another way, the first great step for mankind was the first bipedal step taken by our remote ancestors. The moment we started walking on our hind legs and freed our front legs to become our grasping, manipulative, tool-making hands, we were ready to conquer the world.

How long ago did this happen? A new discovery in Ethiopia, announced in 2006, of the fossilised skeleton of a child that could walk upright at the age of three, has revealed that we have been bipedal for at least 3.3 million years. Intriguingly, the little girl, called Selam by her finders, had what has been loosely described as the lower half of a human and the upper half of an ape. In other words, although she had bipedal feet, she had chimp-like hands. This suggests that she spent part of her time walking upright on the ground, like a human, but then, when danger threatened, she took to the trees, using her arms like an ape.

What this means is that, three million years ago, our ancestors' feet were more advanced than their hands. To put it another way, our feet were in the forefront of the evolutionary trend that led to the fully human condition, rather than lagging behind it. We did not develop our bipedally walking feet because our hands had already become specialised as precision grippers, but rather the reverse. Our front feet were able to develop into sophisticated hands because our feet had already taken on the full burden of ground-level locomotion. All we had to do to move to the next chapter of our evolutionary story was to stop clambering up into the trees when we were frightened. We could then give up our climbing hands and develop the manipulating hands that would eventually give us advanced implements and weapons, and a fully human existence.

This still leaves the major mystery as to why little Selam started walking around vertically on her hind legs, while other apes went on all fours. It was such an unusual step to take, literally, that there must have been some very special local environmental pressure to provoke it. We may never discover what that pressure was, but, thanks to Selam, we do know that her feet were crucial to the evolutionary process.

For what it is worth, a guess as to why we stood up on our hind legs is that our ancestors may have been forced into becoming wading apes in a wooded environment where there was a lot of shallow water. Our relatives, the great apes, cannot swim. Their bodies have the wrong proportions and the wrong balance for it, and it is likely that Selam could not swim either. But her kind could have solved the problem of how to move through the water by descending to ground level, wading through the shallow water, arms aloft, and then clambering up into the trees again. This is just about the only explanation for her curious anatomical mixture of lower-half human and upper-half ape.

Whether this is the true picture of what happened three million years ago or not, it is clear that we owe a great debt to our feet and should revere them as one of the most important parts of our anatomy. Perversely, we fail to do this. Instead we abuse them horribly. We sentence them to spend two-thirds of their life inside cramped leather cells. We force them to walk on hard, tiring surfaces, and we completely ignore their health and wellbeing until they are in serious trouble and sending out pain signals we can no longer ignore.

The reason we look down on them, metaphorically, is that we look down on them physically. They are too far away from our specialised sense organs. If we could examine them as closely as we study our hands, we would take more care of them; but they are at the far end of the body, and most of the time they hardly rate a passing thought.

This attitude is fostered by the feeling that damage to our feet cannot be lethal. It is an undeniable fact that the foot is not a vital organ like the heart, lungs or liver; but badly treated feet can shorten a life span just as surely as a heart attack. To understand why this should be it is necessary to make some field observations on the walking behaviour of the elderly. Those who have thoughtlessly abused their feet for decades are finding old age a time for hobbling along painfully at snail's pace. Others who still have efficient walking feet can stride out for long constitutionals. Taking long walks in old age has emerged as one of the best life-stretchers there is. A survey of exceptional individuals who live into their nineties and beyond reveals that a remarkably high proportion of them have been devoted walkers, often covering several miles a day, every day. There is something about relaxed walking that exercises the whole body in an ideal way. The currently fashionable habit of jogging, on the other hand, can cause all kinds of problems, except for younger adults. The feet love gentle movement; they hate jarring movement.

Every time the foot touches the ground as we move forward, no matter how soft the step, it receives a jolt. In an average life of moderate activity, the feet hit the ground millions of times. The first moment of each contact consists of the heel pad slamming down and acting as a shock absorber. We take this vital action completely for granted, but it is only necessary to miss a step up or a step down in the dark to realise how unpleasant it is to touch the ground in some other, unprepared way.

The split second following this initial contact, the foot has switched roles. From being a shock absorber it now becomes a rigid support structure for the moving weight of the body. Finally, through its toes, it becomes a pushing organ projecting the body forward. This triple sequence occurs with every step we take.

To make all this possible, the foot is a remarkably complex structure. It contains 26 bones, 33 joints, 114 ligaments and 20 muscles. Leonardo da Vinci called it a masterpiece of engineering, and, when you consider the special kind of balancing trick it must perform for our uniquely upright bodies, you are forced to agree with him. Imagine, for instance, a solid, life-size dummy of a standing human being, with the weight distributed in a natural fashion, and consider what would happen if it were to be given a gentle push. It would crash over immediately, a top-heavy disaster. Imagine what would happen if such an object were placed on the side of a hill or on some kind of sloping ground. It would topple over instantly. Despite this we are remarkably nimble. This is because the feet are sending and receiving countless messages during every second of human movement, resulting in thousands of minor muscle adjustments enabling us to keep a balanced view of the world. Even when we are standing still and seemingly inactive, the feet are busily working away, making tiny, subtle and almost imperceptible alterations in our posture.

To achieve all this we have had to make one special sacrifice during the course of evolution. As one anatomist put it rather colourfully, we had to develop webbed feet. What he meant by this was that our big toes had to give up their opposability and become welded to the other toes. In technical terms, this meant that the transverse metatarsal ligament had to spread across all five toes instead of only four. In apes, the metatarsal bones of the big toe are free from the rest of the foot and this makes for a much longer, more prehensile big toe. In humans the five toes are all shorter and more tightly webbed together. We can still wiggle our toes, but we have lost our natural ability to grasp with them.

One ability we have not lost is leaving scent signals with our feet. It is claimed that Australian Aborigines have been able to tell the identity of individuals by smelling their footprints some time after they have passed by. In such cases, of course, the walkers in question were barefoot, but dogs can track human footprints even when the individuals concerned have been wearing thick shoes. This is possible because the soles of the feet are more richly supplied with sweat glands than any other part of the body except the palms of the hands, which are themselves ex-feet, of course. These sweat glands are highly susceptible to stress and they increase output dramatically whenever we are under pressure. We become aware of our sweaty palms, but we do not always realise that our feet are following suit. The scent produced is so strong that enough of it can leak out through our socks and shoes to leave a scent trail that, even when two weeks old, is child's play for any bloodhound's nose. It seems highly probably that in our primeval, barefoot past, when our species was much thinner on the ground, our foot-scent signals were of some considerable importance to us in keeping tabs on both our friends and our foes.

Today the odour-producing ability of our feet has become nothing but a nuisance. Only toiletry manufacturers benefit by it. Because the scented sweat is trapped inside the prison of socks and shoes, it quickly falls prey to bacterial action and goes stale.

Another property of our feet that has largely lost its use is the ridging of the ventral skin. We have toeprints that are every bit as individual as our fingerprints and which could be used in the same way to identify us. Their original, anti-slip function has become almost meaningless in cultures where the wearing of shoes is the norm.

This ridged skin on the soles of the feet and the palms of the hand is called *volar* skin and it has one very strange property. It never gets sun-tanned. The obvious response to such a statement is to point out that these are the two areas that are usually hidden from the sun, but this is not the true answer. If the palms and soles are deliberately exposed to the sun, they still remain untanned. Something in the human body specifically inhibits the production of additional melanin in these regions, leaving the palms and the soles a paler colour than the rest of the sun-tanned body. Even very dark-skinned races still retain pale soles and palms, suggesting that this quality is part of the evolutionary heritage of our whole species. The explanation put forward by those who have studied this phenomenon is that hand and foot gestures are made more conspicuous by this device. With hand gestures this is easy to accept, but it may seem rather far-fetched to suggest that sole-of-the-foot gestures were ever that important. The examination of foot movements in moments of emotional conflict, which follows later, will make this less difficult to understand.

In tribal cultures where it is still the norm to go barefoot, it is possible to see the strength that can be developed in this lowest part of the human body. The Samoan Fuatai Solo climbing a coconut tree has to be seen to be believed. He has been known to shin barefoot up a 30-foot tree-trunk in less than five seconds. By comparison all well-heeled urbanites come into the category which Western ranchers used to refer to as tenderfoot.

Fuatai Solo set up his barefoot record in Fiji in 1980, and it is on the Fijian islands that one can still observe an even more incredible achievement of the human sole: fire-walking. The event begins with a long period of prostrate relaxation. The fire-walkers, who will perform in the evening, gather and lie down quietly together for several hours. Then, after dark, they light logs of wood in the fire pit and begin to heat the large smooth stones that lie packed tightly together beneath the burning branches. These are huge pebbles collected from the shore, and when the fire has reached such a heat that they are glowing, the fire-walkers start to rake away all the embers. They do this until only the pebbles are left, fully exposed and still hot enough to ignite a handkerchief if one is dropped on them. At this point, these extraordinary men walk, in their bare feet, across the baking pebbles as if crossing a shallow river on stepping-stones.

Logic demands that the soles of their feet should be seriously blistered, that the flesh of their feet should be cooked, but they remain unharmed. I have personally examined the soles of the feet of these men immediately after their performance and, to my surprise, found that they were unusually soft and spongy. They were certainly not calloused or horny, nor were they secretly treated in any way. I also examined the stones in the fire pit and found they were still intensely hot the *morning after* the fire-walk. I have no explanation for this amazing achievement of the human foot. Other investigators have found themselves similarly baffled, and the explanations put forward seem grossly inadequate. The best attempt suggests that when the skin comes into contact with intensely hot surfaces the natural moisture of the body vaporises so fast that it forms a protective layer between the skin and the stone. Although it is just possible to conceive of such a process, and to imagine a sort of human hovercraft walking along on a thin cushion of rapidly expanding vapour, the whole idea collapses when one recalls the last time one touched a hot stove, screamed with pain and suffered a severe blister. For the moment, the fire-walking ability of the male foot must remain a fascinating enigma.

THE FEET

When we are born our tiny feet are soft and floppy, and about one-third of the adult length. It takes them twenty years to complete their long, slow growth, and it is a mistake to rush them. Eager parents who try to persuade their offspring to walk before they are ready to do so may actually cause harm to their feet. Even more damaging are tightly restricting bedclothes. These may tuck the baby in snugly, but if the sheet is too taut over the lower legs the softly pliable feet may be twisted and squashed while the infant sleeps. Stiff, constricting shoes and clinging socks can also compress the soft infant foot and all these impositions may, in extreme cases, stretch young ligaments out of shape and throw the soft bones out of alignment.

Further damage is done to the growing feet during schooldays, when parents delay replacing outgrown shoes. Tight shoes or boots crush the toes and cause permanent damage. This enemy of the human foot has been with us for centuries and shows little sign of retreating, despite our knowledge of the trouble it causes.

Fortunately for the human male, foot size shows a marked gender difference, with the male foot being both longer and broader than its female counterpart. So it follows that any gender signal exaggeration will damage a male's foot less than that of his female partner. If a super-female foot is small, dainty and horribly crushed, a super-male foot will be big, broad and generous. In most cultures, therefore, the exaggerated male foot and its footwear will show harmless increases instead of harmful reductions. Furthermore, male foot-comfort will be aided by the bizarre but widely held belief that unusually large feet in a man mean that he has a large penis.

This belief may explain the development in medieval Europe of men's shoes so long that they made it difficult to walk. Shoes with very long toes were first worn in Western Europe as early as the twelfth century. As often happens with a new male fashion, they started out as a cover-up for a powerful man. In this case it was the Count Fulk of Anjou who was said to suffer from a foot deformity that forced him to wear overlong shoes.

These pointed shoes where known as 'pigaches' and they extended about 2 inches beyond the toes. To prevent the point from drooping they were stuffed with wool, moss or hair. Some of them ended decoratively in a fishtail-, serpent- or scorpion-shaped tip, but these extravagances were confined to only the most noble of feet.

In the fourteenth century, European merchants discovered an even longer shoe in Poland and brought it back to the West where the pointed toe section was referred to as the 'poulaine'. The shoes were also called 'pikes', and their pointed part now extended at least 4 inches beyond the toes.

Some of these wildly exaggerated shoes were so cumbersome that they had to be supported by a cord or chain that ran from their tip up to the man's knee. In an early *Survey of London* there is a reference to the fact that the fashionable men of the city used to wear 'piked shoes, tied to their knees with silken laces, or chains of silver or gilt . . .'. They made fast walking so difficult that they sometimes became a liability. On one important occasion, at least, they caused the death of the wearer. Duke Leopold II of Austria died when his poulaines prevented him from escaping pursuing assassins. Knights in battle were known to cut off the pointed tips of their shoes when they dismounted to fight on foot.

The fashion seems to have reached ridiculous lengths, literally, during the fourteenth century, with poulaines that had 'pointed toes that curled at the end . . . ranging from six inches for a commoner, up to two feet for a prince'. As the fashion continued to spread, the nobles became irritated that the lower orders were beginning to copy them more and more, and eventually the King, Edward III, had to introduce a new law to stop the rot. This stated: 'No Knight under the estate of a lord, esquire or gentleman, nor any other person, shall wear any shoes or boots having spikes or points exceeding the length of two inches, under forfeiture of forty pence.'

The stuffed, often upturned points of these ridiculous shoes had a distinctly phallic look and were clearly meant to act as a sexual display. Some of the shoes even had images of male genitals painted on them, to emphasise their erotic significance. Others were painted flesh-pink to make their stiffly erect tips more sexually explicit. The lining of the shoes was sometimes made from soft fur designed to look like pubic hair. When the feet moved the erect tips bobbed up and down in a suggestive manner and playing 'footsie' under the table took on a whole new intensity. There was even a poulaine code, in which a wearer who was on the hunt for a girl would have small bells attached to the tips of his shoes to signal his intentions. Young men would stand on street corners and waggle their shoes at passing women.

The Church became increasingly alarmed by this trend and, when it emerged that fashionable young men were being prevented from kneeling properly in prayer, the Vatican decided to take action. It pronounced men's fashion shoes as wicked, and a disgusting example of declining morality in male society.

Inevitably this made them even more popular and the fashion refused to go away. It was not until the late fifteenth century that they finally disappeared, and that had nothing to do with the Church. It was simply that they had become suddenly unfashionable and were being replaced by the blunt 'cowmouth', 'duck's bill' or 'bear paw' shoe. This also exaggerated the larger size of the male foot, but it did so in width rather than length. In their most exaggerated form they were said to be as much as 12 inches (30 cm) wide, forcing men to progress in a strange, waddling gait.

As with the pointed, phallic shoe, the broad shoe began life in response to an anatomical defect of a dominant male. The French King, Charles VIII, suffered from polydactylism – he had six toes on each foot – and he needed an unusually wide shoe to contain them.

In recent times, men's shoes have generally become more practical in design and have been far less vulnerable to extremes of fashion. One exception to this rule was the winklepicker shoe of the late 1950s and early 1960s. Like the early piked shoes, they had long, pointed toes, but, unlike them, the points were sharp. They were less phallic and more like a spearheaded weapon. Indeed, they were especially popular among young gang members who used them to kick their fallen enemies. In gang fights, if a rival had collapsed on the ground, well-placed jabs with a sharp-pointed winklepicker shoe could cause serious injury, especially to the eyes and testicles.

Eventually, as so often happens with extreme fashions, the pointed toes of these shoes became so long that they made walking difficult and the winklepicker was replaced by chisel-toe shoes in the

1960s. They have resurfaced today, in a minor way, in the modern goth cult, where they have reverted to their earlier title of pikes.

Another modern trend towards extreme male footwear is the development of a much heavier shoe, sometimes called the Stomping Shoe. This is a shoe worn by aggressive young men with an urge to kick or stamp on their enemies. It surfaced in the 1950s as the crepe-soled Beetle Crusher of the English Teddy Boys, in the 1960s as Desert Boots, in the 1970s as Timberland Boots and in the 1990s as the sturdy Doc Martens. Such boots not only act as valuable weapons during street fights, but also transmit a visual signal of heavy-footed male hostility. They are popular today among young gang members and football hooligans, but their origins can be traced back to early Egypt where, amusingly, some young men used to wear heavy sandals with pictures of their enemies painted on the soles.

In ancient times men's shoes sometimes took on a special significance that is hard for us to understand today, when wearing shoes is commonplace. In early civilisations, shoes could stand for liberty because slaves went barefoot. The removal of shoes in certain places of worship presumably arose as an act of humility in the presence of the deity, for whom the worshipper was a willing slave.

The feet themselves, in ancient times, were often seen as the site of the human soul and it has been pointed out that in Greek legend lameness indicates some defect of the spirit or moral blemish. An even more ancient symbolism sees the feet as the rays of the sun, with the swastika symbol arising as a sun-wheel with feet.

Turning from foot symbolism to the body language of the feet, an interesting fact emerges. The feet are undeniably the most honest part of the whole human body. Small movements and posture shifts of the feet tell us the truth about a person's mood. The reason for this is that we seldom think what we are doing with our feet. When we meet other people we concentrate on their faces and we know that they concentrate on ours, so we become proficient liars with our smiles and our frowns. We put on the face we want others to see. But as you travel down the body, away from the facial region, the body language becomes progressively more sincere. Our hands are about halfway down and they are halfway honest. We are only vaguely aware of their actions, but we can lie with them to some extent. However, the feet, at the other end of the body from the crucial facial region, are left to their own devices, and that is why they are so worthwhile studying.

The man who sits in his chair being interviewed looks so calm and relaxed. He is smiling softly and his shoulders are unhunched. His hands make smooth, gentle gestures. He appears completely at ease. But look now at his feet. They are wrapped tightly round one another as if clinging to each other for safety. Now he unwinds them and almost imperceptibly starts tapping one foot on the ground, as if he is trying to run away without moving. Finally, he crosses his legs and the foot that now hangs in the air starts to flap up and down, again trying to flee while he stays on the spot. One famous interviewer in New York used to perform the foot-flap with such regularity that his television colleagues made a close study of it. They found that it only occurred when he was not at ease with his guest and they suggested that if he had the word HELP written on the sole of his shoe, the significance of his foot signal might become nationally understood.

Sometimes, the foot-tapping action of impatience, with its urgent signal of the desire to flee, becomes reduced to no more than a toe wiggle, with the toes being raised and lowered almost invisibly. Like all foot-shifting and foot-jiggling movements these actions represent a suppressed desire to walk or run away from the situation being faced by the foot-mover. Lecturers, who would often like to flee from their audiences, perform a whole range of foot actions indicating their true mood. At a long conference, a study of foot movements of speakers is often more interesting than what they are saying. Unfortunately, it has become common practice for conference organisers to hide their speakers' bodies behind a lectern or some other barrier that prevents direct observation of their truthful lower appendages. Lecturer specialities, when they can be seen, include a delightful range of heel-raisings, foot-teeterings, swayings, pacings and tappings, as if the feet were trying every possible way of making their escape from the hostile stare of hundreds of pairs of eyes.

When a man is sitting down with his legs crossed, boredom is often expressed with yet another foot action, the mid-air multi-kick. This has a slightly more hostile tone to it than the foot-flap, as if the performer wishes to kick whoever is causing the boredom. The crossed-over leg repeatedly kicks forward into the air, but travels only a very short distance before returning to its original position. This is only a short step away from the full foot-stamp of anger.

Foot-shuffling and shoe-scuffing belong to a slightly different mood, typified by the small boy who is being quizzed about some misdeed. In his case, the movements of the feet lack the urgent, rhythmic beat of the multi-kicker or the foot-tapper. The boy's feet are not signalling outright attack or flight, or even marching defiantly off into the distance. Instead their irregular twists and shifts show that what they really want to do is quietly sneak away.

Interpersonal contacts involving feet are few and far between, except for professionals such as chiropodists and masseurs. Lovers exploring each other's bodies may kiss each other's feet and toes, but for most people this plays a very minor role in their sex lives. Equally rare today is that other form of foot-kiss, the humble act of lowering the body to kiss the lowest part of a dominant figure. This is an act of extreme submission and subordination of a kind that is alien to modern societies. In ancient times, when rulers enjoyed a loftier status, the lips of the lowly and the feet of the lordly met on many an occasion. Diocletian, the Roman emperor who ruled as an absolute monarch, insisted that senators and other dignitaries kiss his foot when he received them and again when they left his presence. Even relatives of Roman emperors had to kiss the Imperial Foot. Today such activities are restricted largely to foot fetishists, who pay specialised prostitutes large sums of money to be allowed to grovel at their feet.

The opposite side of the coin from grovelling at someone's feet is placing your foot on top of someone. This is an act of dominance that has been formalised in a number of contexts, the most familiar of which is the brave hunter who, having gunned down some innocent wild beast, stands proudly by his kill with gun in hand and one foot planted firmly on the back of the corpse. An old Polish Jewish custom involved treading on the foot of the spouse at the wedding ceremony. Whichever partner managed to do this first was supposedly destined to become the dominant member of the marriage. For upper foot read upper hand.

Although many of these old customs are dying out today, in Scotland the ceremony of the First Foot still survives. In this, good fortune for a coming year depends on the arrival on your doorstep of the First Foot of the New Year, no more than a few minutes after midnight as 31 December becomes 1 January. The newcomer must be carrying gifts and, if the household is to prosper for the next twelve months, he must be a dark male stranger who does not suffer from flat feet. It is vital that he enters the house with his right foot, as the left foot is considered extremely unlucky.

In earlier days, all persons entering a house had to ensure that it was their right foot that was placed first over the threshold. Grand residences employed special servants whose task was to ensure that nobody forgot this procedure. The reason for this preoccupation with left and right feet was that, as with other left and right distinctions, God was thought to operate through the right foot and the Devil through the left. To put one's best foot forward was to stride out with the right foot. The right foot was good and kind; the left foot was wicked and hostile. This, incidentally, is why armies generally put the left foot forward first when setting off on a march. A typical order is 'Quick march! Left, right, left, right.' The hostile left foot is deliberately moved first to show the hostile intent of the marching men, although it is doubtful whether many modern-day soldiers are aware of this small piece of superstitious military history.

Finally, a word about the name 'football'. It has always been assumed that this title refers to the fact that the world's favourite game involves kicking a ball with the foot. This is so obvious that nobody has ever questioned it, but it happens to be wrong. Football has been played in Britain for nearly a thousand years and, for most of that time, the male foot was not much involved. The early form of folk football was a rough and ready sport, played by large crowds who wrestled for the ball and, once gaining it, held on to it as tightly as possible. To have kicked it into the milling crowd would have lost it. This handling game was so rowdy that it was repeatedly banned, but refused to disappear. It can still be seen in a few places, even today. At Ashbourne in Derbyshire the medieval game is played each year on Shrove Tuesday and Ash Wednesday, when up to two thousand players assemble in the town centre. The ball is ceremonially thrown into the crowd at two o'clock and the huge scrum then fight for it and try to get it to one of the two goals, which are several miles apart. If nobody has succeeded by late in the evening, the game is called off.

That type of football survived right up until the nineteenth century, when some of the top public schools started kicking the ball instead of running with it. In 1863 the kicking game was given specific rules that outlawed handling and the modern game of soccer was born. The handling game still survived, as rugby, and was eventually exported as Australian Rules and American Football. But it was soccer, the 'kicking only' game, that went on to become the dominant form of ball sport all over the world.

So, the question remains: why did a game in which the ball was held in the arms rather than kicked with the foot acquire the name of football, centuries before it truly became a game played with the foot? The answer is that early football was so called, not because it was played *with the foot*, but because it was played *on foot*. It was the game of the common people who could not afford the more fancy games that were played on horseback.

Today, a star footballer's foot is a piece of male anatomy on which millions are spent by the top clubs. Real Madrid paid Juventus £44 million for the French player Zinedine Zidane. And the players themselves often insure their precious feet against the risk of serious injury. Some years ago a Brazilian player had his left foot insured for a million pounds, but that sum has long been surpassed. If, during his time at Real Madrid, the Portuguese footballer Luis Figo suffered serious damage to his feet that would see an end to his career, an English insurance company agreed to pay him the staggering sum of £40 million. The foot may be the lowest part of the male body, but for some it is also of the highest importance.

One sports official who had an entirely different attitude to the precious feet of star athletes and footballers was Saddam Hussein's extraordinary son Uday. Having elected himself the head of both Iraq's Olympic Committee and its national football team, he threatened to punish any man who disappointed him. He is reputed to have kept a private scorecard, with written instructions on how many times each player should be beaten on the soles of his feet after a poor performance. For some strange reason, this unusual form of team motivation failed in a spectacular fashion.

24. THE PREFERENCES

It would be wrong to end this book without a comment on the sexual preferences of the human male, as this has become a hotly debated topic in recent years. There are four lifestyles with which a man can occupy his body during his brief span on earth: heterosexual, bisexual, homosexual or celibate.

Viewed purely from an evolutionary standpoint, there is only one valid biological lifestyle for the human male and that is heterosexual. Like all higher forms of life, the human species relies on sexual reproduction to avoid extinction. If a man does not allow his sperm to fertilise an egg at least once during his lifetime he has no chance of passing his genes on to the next generation, and the genetic line, hundreds of millions of years long, that led up to his appearance on earth is terminated.

When our species was young and there were very few humans on earth, fertility must have been a crucial issue. Anything interfering with reproductive success would have been extremely damaging. However, as our numbers grew the situation changed, and when we reached a state approaching human overpopulation, rapid breeding not only became less important, but even dangerous. In the last forty years the global human population has risen from 3,000 million to over 6,000 million. If it continues at this rate the time will come, during the next few centuries, when extreme overcrowding will bring an end to our species. Like a great plague of human locusts we will have devastated the planet.

It follows from this that, today, any adult male who chooses not to breed during his adult life will be helping to slow down the human population increase. This means that monks, priests, eunuchs, bachelors, other celibates and homosexuals are all useful non-contributors to the population explosion. Long ago they would have been wasted breeding units, but today they are valuable non-breeders. This probably explains why, in most advanced countries, where issues of human overpopulation are already widely under-stood, the laws against male homosexuality have recently been relaxed or abandoned. If two men wish to live together as a couple and deny themselves the genetic fulfilment of becoming fathers, they are doing the human species a favour, and Western society is increas-ingly happy for them to do so.

Officially, of course, other reasons are given, such as human rights, privacy laws, sexual liberation, and the rest. But the truth is that, when society makes a major shift in its attitude towards some basic pattern of human behaviour there is usually an under-lying factor at work, a factor that has to do with the biological rules of life.

It should be said, however, that the change in attitude to the homosexual lifestyle is far from global. No fewer than seventy-four countries still have laws against it, with penalties ranging from death to one year in prison. In Afghanistan, Iran, Iraq, Mauritania, Saudi Arabia, Sudan, Yemen and the Islamic regions of Nigeria any adult male caught committing a homosexual act is put to death.

It is no accident that these are all Islam-dominated states. Islamic teaching is stridently opposed to homosexual acts and some of the punishments meted out are bizarre in the extreme. In Afghanistan, for example, a condemned homosexual is executed, either by being thrown off a high roof or hill, or by being buried beside a wall that is then toppled over him. In Iran, the condemned man is given a choice. He can be hanged, stoned to death, halved by a sword or dropped from a high place.

Islam is not alone in outlawing homosexual acts. All the other major religions – Christianity, Judaism, Hinduism – are officially opposed to them. Christianity and Judaism both take their lead from the Bible where it states that a man shall not 'lie with another man as with a woman', describing this as an abomination. In India the maximum penalty a judge can impose for a homosexual act is life imprisonment. Even controversial faiths such as Scientology are opposed. Its founder, the science fiction writer L. Ron Hubbard, called homosexuals dangerous, physically ill, sexual perverts.

In modern times those individuals who wish to adopt a less bigoted approach to human sexual conduct, but at the same time feel themselves compelled to follow the rules of a major faith, have found themselves in a state of acute conflict. When a man is opposed to a categorical statement made by his religion it requires a great deal of double-talk and fancy waffling to reconcile common-sense beliefs and strict religious beliefs.

Returning to a biological approach to human sexual preferences, it is clear that, if we are not crying out for more breeding males, then the existence of homosexual males is no disadvantage to modern human society. Their isolation as a special case in an era dominated by scientific thought is unjustified. As a result, the general view is that, providing they are not, as one Edwardian actress famously put it, 'doing it in the street and frightening the horses', then what happens in the privacy of the home between consenting adults is viewed as their business and nobody else's.

The sexual acts performed by homosexuals are not unknown to heterosexual couples and, religious bigotry apart, it is difficult to understand why these men have aroused so much hostility in the past. It is not as though they belong to a group given to violence. Even their most fanatical supporters have never behaved as badly as the religious fanatics who oppose them, and who have been guilty of everything from witch-hunting and widow-burning to penis-mutilating and suicide bombing.

The question remains as to why a certain, small percentage of adult human males, with or without the approval of society at large, find members of their own gender attractive as sexual partners. Evolution has gone to a great deal of trouble to ensure that it is the opposite sex that is erotically appealing, so how can it be that so many men have somehow switched off these basic responses?

When questioned about the onset of their same-sex interest, many homosexuals state that from boyhood onwards they felt a strong attraction to other males, and never felt drawn to young females. This sets them apart from young boys who often play homosexual games with their male friends, but who then pass on to a new phase, when their interest switches to girls. For the lifelong homosexuals this switch never happens. To understand why this might be, it is important to look at the typical sequence of events that occurs in the first twenty years of the life of the human male.

For the first few years toddlers make no distinction between male and female friends. Then, when they reach the age of four or five the sexes suddenly draw apart. For a small boy, the little girls who were his close friends only a few weeks before must now be avoided. Now he must play only with other boys. Nobody tells him to do this, it just happens. He becomes part of a group or gang and the boys hang out together. This phase will last about ten years, during which time he will be going through an intensive educational period, programming the amazing computer inside his skull. Even if boys and girls go to school together during this phase, they will separate from one another socially. Indeed, despite modern educational theory, mixing boys and girls during this phase of growth is of little advantage. It may even be distracting.

This ten-year learning phase is something that other primates do not have. They reach sexual maturity in about half the time but, of course, they have smaller brains and far less to learn. The boys-together schooling phase is something special that has been added to the human life cycle. At the end of it, in the early teens, the bodies of both boys and girls start to flood with sex hormones and now, suddenly, the opposite sex is of interest again. During the ten-year stand-off they have become distant objects, often disliked. Now they are a new shape and have new features, as the secondary sexual characters begin to develop.

So the stand-off period has made the opposite sex into a novelty, a mystery, something to be explored. (For boys, this reaction does not apply to their sisters, of course, because as siblings they have been pushed close together by family constraints, a fact that helps to avoid incest.) At this point boy-meets-girl is a theme that dominates the lives of teenagers, and intense sexual exploration is not far away. There will be a brief period when there is a conflict between the old, all-boy gang and the new interest in girls. Each boy will have to report back to his chums to tell them how he has progressed with a particular girl, until, one day, there is a stubborn refusal to give them any details, and they know instantly that they have lost one of their group.

Returning now to the boys who do not reach the teenage heterosexual phase, what happens is that for some reason they get stuck in the stand-off phase, and stay there for the rest of their lives. They cannot understand why young boys, who were playing sex games with them only a few months before, are now only interested in chasing girls. The all-boy phase seems perfect and when sexual maturity arrives, they feel no urge to abandon their all-male social existence. Their sex hormones activate them erotically, but their focus of interest is still masculine. This is how the lifelong homosexual male starts his sexual journey, but why does it happen to just a few boys while the majority move easily to the heterosexual phase?

The answer seems to be that it is the unique addition of such a lengthy ten-year learning phase in our species that causes the problem. During that phase, male bonding is intense and male-to-male attachment is powerful. It takes a massive jolt from the sex hormones at puberty to break down the boy-to-boy loyalties, and if there are any special social factors adding their weight at this point, the break can be thwarted. These factors can be of several kinds. A boy who has especially unpleasant experiences with girls during the stand-off phase may find that, even flooded with sex hormones, he cannot switch into the state where he finds them appealing. Or, he may have found the boyish sex games that are so common in the stand-off phase to be particularly exciting and this may have fixated him on other males as sexual companions. For him it is impossible to make the switch at puberty because he cannot bear to leave behind what he had before.

There are many other social factors that are impinging upon the pre-pubertal male and imprinting upon him powerful attachments. The reason it happens to him and not to young chimpanzees or young monkeys is that other species lack this vital stand-off phase and are never put in this position of key switch from boys-together to boy-plus-girl.

In his study of what he calls *The Eternal Child*, zoologist Clive

Bromhall has put forward the idea that this extended childhood is part of a general infantilising of the human species, a process he sees as the basis of our evolutionary success story. As a way of maximising our human playfulness and curiosity, evolution has made us more and more childlike over the past million years or so. While this has made us more inventive and given us the technology that has made us great, it has also had certain side effects. To explain these Bromhall suggests that there are four types of human male.

There is the Alphatype, who is the least juvenile male. He is like an alpha male ape, ruthless, determined, ambitious, strong and intolerant. Then there is the Bureautype, still concerned with high status, but much more cooperative, making him the perfect business partner. Thirdly there is the Neotype, more childlike, the exuberant, fun-loving family man. And finally there is the Ultratype, imaginative, insecure, and unable to move on past the all-boy phase of childhood.

Bromhall sees this last type, to which homosexual males belong, as merely a by-product of the increasingly infantile condition of our species. In other words, when evolution took the human species down the road of increasingly playful, innovative behaviour as a new survival device, the process was not too precise. The ideal outcome would have been to create a species made up of a balanced mixture of reliable organisers, the Bureautypes, and creative fun-lovers, the Neotypes. But this shift was not fully achieved. At one extreme there remained a few of the old-style Alphatypes, the macho toughs, good in a fight, but poor cooperators; and at the other end of the scale a few of the new-style Ultratypes, so advanced in this new evolutionary direction that they got stuck at the boy-group stage.

If, as a result of this, the Ultratypes accidentally became 'reproductively challenged' they also became unusually imaginative and intellectually inquisitive. Bromhall reports that their academic achievements are well above average. A male homosexual is six times as likely to gain a college education and sixteen times as likely to have a Ph.D. as males in general.

But what of the future? The fact is that people deserve to be

257

treated as individuals and respected for their personal merits rather than as members of a group which they did not join but which was thrust upon them. Isolating homosexuals as though they are members of some exclusive club does them no favours. It encourages bigots to attack them, which makes about as much sense as outlawing left-handers or redheads.

This general rule also applies to the male body as a whole. Throughout this book, we have seen how the basic human blueprint has varied in many ways and how men have repeatedly tried to alter their particular version of it. The very tall have wanted to be shorter and the very short have wanted to be taller; the very fat have wanted to be slimmer and the very thin have wanted to put on weight; the straight-haired have wanted waves and the wavy-haired have tried to straighten their hair; and so on. But there is no harm in variation. It is evolution's way of ensuring that, when a dramatic environmental change arrives, there will be somebody, somewhere, who will be better equipped to deal with the new conditions. We should cherish our differences and not try to smother them. And we should put behind us, once and for all, the rigid beliefs and ancient bigotries that demand that we should all think the same way, look the same way and behave the same way. Variety is not just the spice of life, it is the very food of life.

References

THE EVOLUTION
Morris, Desmond. 1997. *The Human Sexes.* Network Books, London.

THE HAIR
Aurand, A. Monroe. 1938. *Little Known Facts about the Witches in our Hair. Curious Lore about the Uses and Abuses of Hair Throughout the World in all Ages.* Aurand Press, Harrisburg, PA.
Berg, Charles. 1951. *The Unconscious Significance of Hair.* Allen & Unwin, London.
Cooper, Wendy. 1971. *Hair: Sex, Society, Symbolism.* Aldus Books, London.
Freddi, Cris. 2003. *Footballers' Haircuts.* Weidenfeld & Nicolson, London.
Macfadden, Bernarr. 1939. *Hair Culture.* Macfadden, New York.
Segrave, Kerry. 1996. *Baldness. A Social History.* McFarland, Jefferson, NC.
Severn, Bill. 1971. *The Long and Short of it. Five Thousand Years of Fun and Fury Over Hair.* David McKay, New York.
Sieber, Roy. 2000. *Hair in African Art and Culture: Status, Symbol and Style.* Prestel Publishing, New York.
Trasko, Mary. 1994. *Daring Do's. A History of Extraordinary Hair.* Flammerion, Paris.
Woodforde, John. 1971. *The Strange Story of False Hair.* Routledge & Kegan Paul, London.
Yates, Paula. 1984. *Blondes. A History From Their Earliest Roots.* Delilah, New York.

Zemler, Charles De. 1939. *Once Over Lightly, the Story of Man and his Hair.* Author.

THE BROW
Cosio, Robyn, and Robin, Cynthia. 2000. *The Eyebrow.* Regan Books, New York.
Lavater, J. C. 1789. *Essays on Physiognomy.* John Murray, London.

THE EARS
Mascetti, Daniela, and Triossi, Amanda. 1999. *Earrings from Antiquity to the Present.* Thames & Hudson, London.

THE EYES
Argyle, Michael, and Cook, Mark. 1976. *Gaze and Mutual Gaze.* Cambridge University Press, Cambridge.
Coss, Richard. 1965. *Mood Provoking Visual Stimuli.* UCLA.
Eden, John. 1978. *The Eye Book.* David & Charles, Newton Abbot.
Elworthy, Frederick Thomas. 1895. *The Evil Eye.* John Murray, London.
Hess, Eckhard H. 1972. *The Tell-Tale Eye.* Van Nostrand Reinhold, New York.
Gifford, Edward S. 1958. *The Evil Eye.* Macmillan, New York.
Maloney, Clarence. 1976. *The Evil Eye.* Colombia University Press, New York.
Potts, Albert M. 1982. *The World's Eye.* University Press of Kentucky, Lexington.
Walls, Gordon Lynn. 1967. *The Vertebrate Eye.* Hafner, New York.

THE NOSE
Glaser, Gabrielle. 2002. *The Nose: A Profile of Sex, Beauty and Survival.* Simon & Schuster, New York.
Gilman, Sander L. 1999. *Making the Body Beautiful. A Cultural History of Aesthetic Surgery.* Princeton University Press, New Jersey.
Stoddard, Michael D. 1990. *The Scented Ape. The Biology and Culture of Human Odour.* Cambridge University Press, Cambridge.

THE MOUTH
Anon. 2000. *Lips in Art.* MQ Publications, London.
Beadnell, C. M. 1942. *The Origin of the Kiss.* Watts & Co., London.

REFERENCES

Blue, Adrianne. 1996. *On Kissing: From the Metaphysical to the Erotic.* Gollancz, London.

Garfield, Sydney. 1971. *Teeth, Teeth, Teeth.* Arlington Books, London.

Huber, Ernst. 1931. *Evolution of Facial Musculature and Facial Expression.* Johns Hopkins Press, Baltimore.

Morris, Hugh. 1977. *The Art of Kissing.* Pan, London.

Perella, Nicholas James. 1969. *The Kiss, Sacred and Profane.* University of California Press, Berkeley.

Phillips, Adam. 1993. *On Kissing, Tickling and Being Bored.* Faber & Faber, London.

Ragas, Meg Cohen, and Kozlowski, Karen. 1978. *Read My Lips: A Cultural History of Lipstick.* Chronicle, San Francisco.

Tabori, Lena. 1991. *Kisses.* Virgin, London.

THE BEARD

Adams, Russell B. 1978. *King C. Gillette, the Man and His Wonderful Shaving Device.* Little, Brown, New York.

Berg, Stephen. 1998. *Shaving.* Four Ways Books, New York.

Bunkin, Helen. 2000. *Beards, Beards, Beards!* Green Street Press, Montgomery, AL.

Dunkling, Leslie, and John Foley. 1990. *The Guinness Book of Beards and Moustaches.* Guinness Publishing, London.

Goldschmidt, E. Ph. 1935. *Apologia De Barbis. A Twelfth-Century Treatise on Beards and Their Moral and Mystical Significance.* University Press, Cambridge.

Krumholz, Phillip. 1987. *History of Shaving and Razors.* Adlibs Pub. Co.

Mitchell, Edwin Valentine. 1930. *Concerning Beards.* Dodd & Mead, New York.

Peterkin, Alan. 2001. *One Thousand Beards: A Cultural History of Facial Hair.* Arsenal Pulp Press, Vancouver.

Pinfold, Wallace G. 1999. *A Closer Shave: Man's Daily Search for Perfection.* Artisan, New York.

Reynolds, Reginald. 1950. *Beards. An Omnium Gatherum.* Allen & Unwin, London.

Reynolds, Reginald. 1976. *Beards: Their Social Standing, Religious Involvements, Decorative Possibilities, and Value Offence and Defence Through the Age.* Doubleday, New York.

THE NECK
Dubin, Lois Sherr. 1995. *The History of Beads.* Thames & Hudson, London.

THE ARMS
Comfort, Alex. 1972. *The Joy of Sex.* Crown, New York.
Friedel, Ricky. 1998. *The Complete Book of Hugs.* Evans, New York.
Hotten, Jon. 2004. *Muscle.* Yellow Jersey Press, London.
Stoddart, Michael D. 1990. *The Scented Ape.* Cambridge University Press, Cambridge.
Watson, Lyall. 2000. *Jacobson's Organ.* Penguin Books, London.

THE HANDS
Gröning, Hans. 1999. *Hände; berühren, begreifen, formen.* Frederking & Thaler, Munich.
Harrison, Ted. 1994. *Stigmata. A Medieval Mystery in a Modern Age.* Penguin Books, London.
Lee, Linda, and Charlton, James. 1980. *The Hand Book.* Prentice-Hall, New Jersey.
Morris, Desmond. 1997. *The Human Sexes.* Network Books, London.
Napier, John. 1980. *Hands.* Allen & Unwin, London.
Sorrell, Walter. 1967. *The Story of the Human Hand.* Weidenfeld & Nicolson, London.
Ward, Anne et al. 1981. *The Ring, from Antiquity to the Twentieth Century.* Thames & Hudson, London.
Wilson, Frank R. 1999. *The Hand.* Vintage Books, New York.

THE CHEST
Darwin, Charles. 1871. *The Descent of Man.* John Murray, London.

THE BELLY
Flugel, J. C. 1930. *The Psychology of Clothes.* Hogarth Press, London.
Fryer, Peter. 1963. *Mrs Grundy. Studies in English Prudery.* Dobson, London.
Laver, James. 1964. *Modesty in Dress.* Heinemann, London.

THE BACK
Draspa, Jenny. 1996. *Bad Backs & Painful Parts.* Whitefriars, Chester.
Inglis, Brian. 1978. *The Book of the Back.* Ebury Press, London.

REFERENCES

THE HIPS

Bulwer, John. 1654. *A View of the People of the Whole World.* William Hunt, London.

THE PUBIC HAIR

Kiefer, Otto. 1934. *Sexual Life in Ancient Rome.* Routledge & Kegan Paul, London.

Licht, Hans. 1932. *Sexual Life in Ancient Greece.* Routledge & Kegan Paul, London.

Manniche, Lise. 1987. *Sexual Life in Ancient Egypt.* Routledge & Kegan Paul, London.

THE PENIS and THE TESTICLES

Allen, M. R. 1967. *Male Cults and Secret Initiations in Melanesia.* Melbourne University Press, Melbourne.

Berkeley, Bud. 1993. *Foreskin: A Closer Look.* Alyson Publications, Los Angeles.

Bertschi. H. 1994. *Die Kondom Story.* VGS, Cologne.

Bigelow, Jim. 1992. *The Joy of Uncircumcizing.* Hourglass Publishing, CA.

Bosch, Vernon. 1970. *Sexual Dimensions: The Fact and Fiction of Genital Size.* Ax Productions, Dayton, OH.

Bryk, Felix. 1934. *Circumcision in Man and Woman: Its History, Psychology and Ethnology.* American Ethnological Press, New York.

Bryk, Felix. 1967. *Sex & Circumcision.* Brandon House, North Hollywood, CA.

Chance, Michael. 1996. 'Reason for the externalization of the testes in mammals.' In *Journal of Zoology* 239, Part 4, pp. 691–5. Zoological Society, London.

Cohen, Joseph. 2004. *The Penis Book.* Broadway Books, New York.

Constans, Gabriel. 2004. *The Penis Dialogues: Handle with Care.* Asian Publishing, Fairfield, CT.

Costa, Caroline de. 2003. *Dick – A Guide to the Penis for Men and Women.* Allen & Unwin, London.

Danielou, Alain. 1995. *The Phallus: Sacred Symbol of Male Creative Power.* Inner Traditions, Rochester, NY.

Denniston, George C., and Milos, Marilyn Jayne. 1997. *Sexual Mutilations: A Human Tragedy.* Plenum Press, New York.

263

Driel, Mels Van. 2001. *The Secret Part: The Natural History of the Penis.* Mandrake, Oxford.

Elgin, Kathleen. 1977. *The Human Body: The Male Reproductive System.* Franklin Watts, London.

Friedman, David M. 2003. *A Mind of Its Own: A Cultural History of the Penis.* Penguin Books, London.

Gore, Margaret. 1997. *The Penis Book: An Owner's Manual for Use, Maintenance, and Repair.* Allen & Unwin, London.

Heiser, Charles B. Jr. 1973. *The Penis Gourd of New Guinea.* S. A. Ann. of the Association of American Geographers, Vol. 63, No. 3.

Kinsey, Alfred, et al. 1948. *Sexual Behavior in the Human Male.* Saunders, Philadelphia.

Knight, Richard Payne, and Wright, Thomas. 1957. *Sexual Symbolism. A History of Phallic Worship.* Comprising: Knight's 1786 *Worship of Priapus,* and Wright's 1866 *Worship of the Generative Powers.* Julian Press, New York.

Masters, William, and Johnson, Virginia. 1966. *Human Sexual Response.* Churchill, London.

Nankin, Philip, and Howard, R. (eds). 1977. *The Testis in Normal and Infertile Men.* Raven Press, New York.

Paley, Maggie. 1999. *The Book of the Penis.* Grove Press, New York.

Paola, Angelo S. 1998. *Under the Fig Leaf: A Comprehensive Guide to the Care and Maintenance of the Penis, Prostate and Related Organs.* Health Information Press, Los Angeles.

Parsons, Alexandra. 1989. *Facts & Phalluses: Hard Facts that Stand Up for Themselves.* Souvenir Press, London.

Payne, Richard. 1894. *A discourse on the worship of Priapus, and its connection with the mystic theology of the ancients. (a new edition) to which is added an essay on the worship of the generative powers during the Middle Ages of Western Europe.* Privately printed, 1865. Reprinted London, 1894.

Purvis, Kenneth. 1992. *The Male Sexual Machine: An Owner's Manual.* St Martin's Press, New York.

Rancour-Laferriere, D. 1979. *Some Semiotic Aspects of the Human Penis.* Bompiani, Milan.

Richards, Dick. 1992. *The Penis.* BabyShoe Publications, Kent.

Ryce-Menuhin, Joel. 1996. *Naked and Erect: Male Sexuality and Feeling.* Chiron Publishers, Wilmette, IL.

REFERENCES

Schwartz, Kit. 1985. *The Male Member: Being a Compendium of Facts, Figures, Foibles, and Anecdotes About the Loving Organ.* St Martin's Press, New York.

Scott, George Ryley. 1966. *Phallic Worship.* Luxor Press, London.

Strage, Mark. 1980. *The Durable Fig Leaf: A Historical, Cultural, Medical, Social, Literary and Iconographic Account of Man's Relations with His Penis.* William Morrow, New York.

Templer, Donald I. 2002. *Is Size Important?* Ceshore Publishing Co., Pittsburg.

Thorn, Mark. 1990. *Taboo No More: The Phallus in Fact, Fantasy and Fiction.* Shapolsky Publishers, New York.

Vanggaard, Thorkil. 1969. *Phallos. A Symbol and its History in the Male World.* Cape, London.

Watters, Greg, and Carroll, Stephen. 2002. *Your Penis: A User's Guide.* Urology Publications.

Wright, Richard, and Wright, Thomas. 1962. *Sexual Symbolism: A History of Phallic Worship.* Julian Press, New York.

THE BUTTOCKS

Aubel, Virginia (ed.). 1984. *More Rear Views.* Putnam, New York.

Hennig, Jean-Luc. 1995. *The Rear View.* Souvenir Press, London.

Jenkins, Christie. 1983. *A Woman Looks at Men's Bums.* Piatkus, Loughton.

Tosches, Nick. 1981. *Rear Views.* Putnam, New York.

THE LEGS

Karan, Donna, et al. 1998. *The Leg.* Thames & Hudson, London.

Platinum. 1990. *Footwork.* Star Distributors, New York. (Described as 'A magazine for foot and leg worshippers'.)

Yarwood, Doreen. 1978. *The Encyclopaedia of World Costume.* Batsford, London.

THE FEET

Anon. 1989. *Foot Steps.* Holly Publications, North Hollywood, CA.

Arnot, Michelle. 1982. *Foot Notes.* Sphere Books, London.

Gaines, Doug (ed.). 1995. *Kiss Foot, Lick Boot: Foot, Sox, Sneaker & Boot Worship.* Leyland Publications, San Francisco.

Vanderlinden, Kathy. 2003. *Foot: A Playful Biography.* Mainstream Publishing, Edinburgh.

GENERAL

Baron-Cohen, Simon. 2003. *The Essential Difference*. Allen Lane, London.

Broby-Johansen, R. 1968. *Body and Clothes*. Faber & Faber, London.

Campbell, Anne (ed.). 1989. *The Opposite Sex: The Complete Guide to the Differences Between the Sexes*. Doubleday & Co., Sydney.

Caplan, Jane (ed.). 2000. *Written on the Body. The Tattoo in European and American History*. Reaktion Books, London.

Cherfas, Jeremy, and John Gribbon. 1985. *The Redundant Male*. Pantheon, New York.

Cole, Shaun. 2000. *'Don We Now Our Gay Apparel'. Gay Men's Dress in the Twentieth Century*. Berg, Oxford.

Comfort, Alex. 1967. *The Anxiety Makers*. Nelson, London.

Comfort, Alex. 1972. *The Joy of Sex*. Crown, New York.

Devine, Elizabeth. 1982. *Appearances. A Complete Guide to Cosmetic Surgery*. Piatkus, Loughton.

Dickinson, Robert Latou. 1949. *Human Sex Anatomy*. Williams & Wilkins, Baltimore.

Ford, Clellan S., and Beach, Frank A. 1952. *Patterns of Sexual Behaviour*. Eyre & Spottiswoode, London.

Fryer, Peter. 1963. *Mrs Grundy. Studies in English Prudery*. Dobson, London.

Ghesquiere, J., et al. (1985) *Human Sexual Dimorphism*. Taylor & Francis, London.

Greenstein, Ben. 1993. *The Fragile Male*. Boxtree, London.

Gröning, Karl. 1997. *Decorated Skin*. Thames & Hudson, London.

Guthrie, R. Dale. 1976. *Body Hot Spots*. Van Nostrand Reinhold, New York.

Katchadourian, Herant A., and Lunde, Donald T. 1975. *Biological Aspects of Human Sexuality*. Holt, Rinehart, Winston, New York.

Kiefer, Otto. 1956. *Sexual Life in Ancient Rome*. Routledge & Kegan Paul, London.

Krafft-Ebing, Richard von. 1946. *Psychopathia Sexualis*. Pioneer, New York.

Lang, Theo. 1971. *The Difference Between a Man and a Woman*. Michael Joseph, London.

Licht, Hans. 1963. *Sexual Life in Ancient Greece*. Routledge & Kegan Paul, London.

Lloyd, Barbara, and Archer, John (eds). 1976. *Exploring Sex Differences*. Academic Press, London.

REFERENCES

Lloyd, Charles W. 1964. *Human Reproduction and Sexual Behaviour.* Kimpton, London.

Maccoby, Eleanor, et al. 1967. *The Development of Sex Differences.* Tavistock, London.

Markun, Leo. n.d. *The Mental Differences Between Men and Women: Neither of the Sexes is to an Important Extent Superior to the Other.* Haldeman-Julius Publications, Girard, KS.

Masters, William. H., and Johnson, Virginia. E. 1966. *Human Sexual Response.* Churchill, London.

Masters, William. H., Johnson, Virginia. E. and Kolodny, Robert C. 1985. *Sex and Human Loving.* Little, Brown, Boston.

Morris, Desmond. 1967. *The Naked Ape.* Cape, London.

Morris, Desmond. 1969. *The Human Zoo.* Cape, London.

Morris, Desmond. 1971. *Intimate Behaviour.* Cape, London.

Morris, Desmond. 1977. *Manwatching.* Cape, London.

Morris, Desmond, et al. 1979. *Gestures.* Cape, London.

Morris, Desmond. 1983. *The Book of Ages.* Cape, London.

Morris, Desmond. 1987. *Bodywatching.* Cape, London.

Morris, Desmond. 1994. *Bodytalk.* Cape, London.

Morris, Desmond. 1994. *The Human Animal.* BBC Books, London.

Morris, Desmond. 1997. *The Human Sexes.* Network Books, London.

Morris, Desmond. 1999. *Body Guards.* Element Books, Shaftesbury.

Morris, Desmond. 2001. *Peoplewatching.* Vintage, London.

Morris, Desmond. 2004. *The Naked Woman.* Cape, London.

Nicholson, John. 1993. *Men and Women. How Different are They?* Oxford University Press, Oxford.

Rilly, Cheryl. 1999. *Great Moments in Sex.* Three Rivers Press, New York.

Robinson, Julian. 1988. *Body Packaging: A Guide to Human Sexual Display.* Elysium, Los Angeles.

Short, R. V., and Balaban E. (eds). 1994. *The Differences Between the Sexes.* Cambridge University Press, Cambridge.

Temple, G., and Darkwood, V. 2002. *The Chap Almanac.* Fourth Estate, London.

Thomas, David. 1993. *Not Guilty: In Defence of the Modern Man.* Weidenfeld & Nicolson, London.

Turner, E. S. 1954. *A History of Courting.* Michael Joseph, London.

Wildebood, Joan. 1973. *The Polite World.* Davis-Poynter, London.

Woodforde, John. 1995. *The History of Vanity.* Alan Sutton, Stroud.

Wykes-Joyce, Max. 1961. *Cosmetics and Adornment: Ancient and Contemporary Usage.* Philosophical Library, New York.
Zack, Richard. 1997. *An Underground Education.* Doubleday, New York.

INDEX

269